TIME-DEPENDENT FRACTURE

Time-Dependent Fracture

Proceedings of the Eleventh Canadian Fracture Conference,
Ottawa, Canada, June 1984

edited by

A.S. KRAUSZ
University of Ottawa
Canada

ORGANIZING COMMITTEE:
B. FAUCHER, *CANMET Dept. of Energy Mines and Resources*
A.S. KRAUSZ, *University of Ottawa*
J.W. PROVAN, *McGill University*
W. WALLACE, *National Research Council of Canada*

1985 **MARTINUS NIJHOFF PUBLISHERS**
a member of the KLUWER ACADEMIC PUBLISHERS GROUP
DORDRECHT / BOSTON / LANCASTER

Distributors

for the United States and Canada: Kluwer Academic Publishers, 190 Old Derby Street, Hingham, MA 02043, USA
for the UK and Ireland: Kluwer Academic Publishers, MTP Press Limited, Falcon House, Queen Square, Lancaster LA1 1RN, UK
for all other countries: Kluwer Academic Publishers Group, Distribution Center, P.O. Box 322, 3300 AH Dordrecht, The Netherlands

Library of Congress Cataloging Card Number: 85-256

ISBN-13: 978-94-010-8748-3   e-ISBN-13: 978-94-009-5085-6
DOI: 10.1007/978-94-009-5085-6

Copyright

TABLE OF CONTENTS

VI

PREFACE

The understanding of time dependent crack propagation processes occupies a central place in the study of fracture. It also encompasses a wide range of conditions: failure under sustained loading in a corrosive environment, fracture under cyclic loading in non-degrading and in corrosive environment, and rupture at high temperature. This list covers probably 90% of the failures that occur in engineering practice.

The process of time dependent fracture is controlled by the physics of atomic interaction changes; it is strongly influenced by the microstructure; and affected by the interaction of the material with the mechanical (load, displacement), the thermal (temperature), and the chemical or radiation environment. To be able to control crack propagation the development of testing methods and the understanding of the industrial environment is essential. The conference was organized in this context.

A call was issued for contributions to the following topics.

THERMAL ACTIVATION. Theoretical papers dealing with the modification of fracture mechanics to accommodate thermally activated processes.

TIME DEPENDENT MICRO-PROCESSES. Presentations covering both the theoretical and observational aspects of creep and fatigue damage in materials whose microstructures may exert a significant influence on crack growth.

INDUSTRIAL APPLICATIONS. Submissions describing the practical application of fracture mechanics and damage tolerance analysis to the determination of useful operating lives.

X

ENVIRONMENTAL EFFECTS.   Papers dealing with engineering materials and/or components exposed to aggressive environments, with and without temperature effects.

The response was gratifying. Leading experts responded; the organizers of the conference are grateful for the large number of excellent contributions.

SUPPORTING AGENCIES

The Natural Sciences and Engineering Research Council of Canada

National Research Council Canada

Energy, Mines and Resources Canada

Ottawa University

McGill University

NUMERICAL MODELLING OF CREEP CRACK GROWTH

S.B. Biner* and D.S. Wilkinson

Institute for Materials Research
McMaster University
Hamilton, Ontario, L8S 4M1, Canada

*Present address:
PMRL-CANMET
568 Booth St.
Ottawa, Ontario

ABSTRACT

In this paper, we summarize some recent results, in which the process of creep crack growth is modelled using time-dependent finite element methods. For stationary cracks, the time-dependent stress-strain field around the crack tip region can be accurately determined. We then proceed to a study of crack propagation. To achieve this a general damage model is developed based on two parameters - a critical effective strain for material failure, and a characteristic propagation distance. The model is used to study crack propagation in $2\frac{1}{4}$ Cr-1 Mo steel. The crack growth rates obtained from these numerical studies agree well with available experimental data for the same material. The correlation of creep crack growth data under non-steady stress fields is also discussed. The results have significant implications for the assessment of structural integrity.

1. INTRODUCTION

There is a well defined need for reliable procedures, with which to assess the remaining life of structural components operating in the creep regime. For thick-walled vessels in particular this requires (as one step in the process) a prediction of the rate of creep crack growth. This is not currently possible in many cases, due in large part, to our lack of understanding of the parameters which control crack growth. Therefore, current procedures often require that a component be retired from service while a significant portion of its life may still remain.

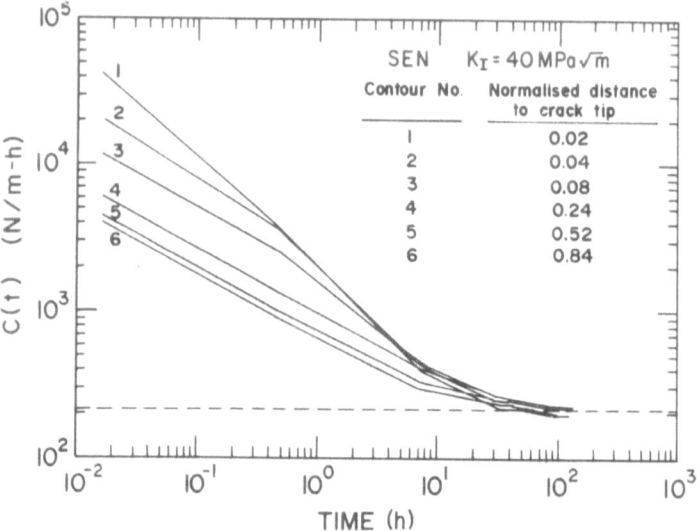

Fig. 1: The variation of C(t) with time, is shown for several integration contours, in the SEN geometry. At long times C(t) is path independent and equal to the tabulated value[3] for C*, as shown by the dashed line.

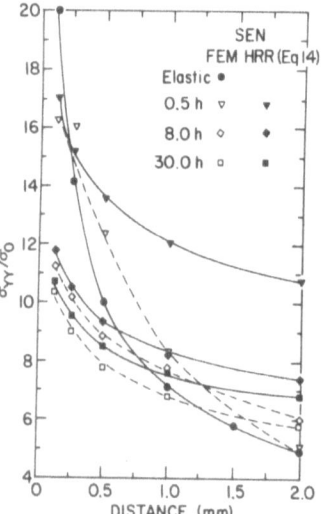

Fig. 2: The stress distribution ahead of the crack at various times is compared with the HRR field using C_B as the amplitude.

A full understanding of creep crack growth requires a synthesis of the macro-mechanics of cracks in creeping solids, and the micromechanisms of creep damage. The first yields information on the time dependent nature of the stress-strain fields ahead of a crack, while the latter yields information on the nucleation, growth and coalescence of cavities in an arbitrary (local) stress field. In this paper we survey our current understanding of these two processes, and develop an approach to combining them in a computational model of creep crack growth.

2. TIME-DEPENDENT STRESS FIELDS

When a crack is initially loaded, the stress field surrounding the crack tip is predominantly elastic. If the material is very brittle, then the elastic stress field continues to dominate throughout the crack growth process and a correlation between crack growth and stress intensity factor K can be expected. However for materials with reasonable ductility, stresses relax considerably ahead of the crack. Eventually, a steady-state field is achieved which, for a power-law creeping material, is expected to exhibit the HRR type singularity. The amplitude of this field is C*, the high temperature analogue of the J-integral.[1]

For many situations of practical interest, the time required to establish this steady-state is comparable to the life of a component or laboratory test. It is thus necessary to study the growth of cracks under non-steady state conditions. We have recently analyzed the development of crack tip fields ahead of a stationary crack using finite element methods.[2] Three specimen geometries have been studied - namely single-edge notched (SEN), double edge notched (DEN), and centre-cracked panel (CCP). For all geometries a crack of length $a/w = 0.5$ is placed under a far-field, mode I load. The load for each geometry is adjusted to produce the same initial elastic stress intensity factor at the crack tip. Other details, including material parameters are given in ref. (2). The time dependent stress fields and the energy rate integrals C(t), are then determined for increasing times until steady-state. The results are shown in Fig. (1) for one geometry. It is clear that eventually, C(t) becomes path independent and converges to the tabulated value[3] for C*. Both the value of C*, and time to reach steady-state t_s, are geometry dependent. These studies have shown that during the transition period, the stress field near the crack tip has the HRR singularity (see Fig. (2)), but that the amplitude is given by

$$C_B(t) = \lim_{p \to 0} C(t) \tag{1}$$

where p is the distance from the crack tip to the path along which C(t) is measured. (Although this definition shows some similarity with Atluri's T_C integral[4], a detailed mathematical analysis of C_B is necessary to show a relationship between these two parameters). C_B has been calculated as a function of time for all geometries. When normalized by the steady-state C*, and plotted against time normalized by t_s, a single curve is found for all geometries. An even more interesting result is obtained if time is normalized by

4

$$t_I = a_m \frac{K^2(1-v^2)}{(m+1)EC^*} \qquad (2)$$

This is the estimated time for the K to C* transition, developed by Reidel and Rice.[1] It is clear from these results (see Fig. (3)), that during the transition period any attempt to correlate crack growth with either of the extreme parameters (K and C*) will fail. However, as will be shown later, crack growth may correlate with C_B since it describes the stress-strain field during both the transition period, and beyond.

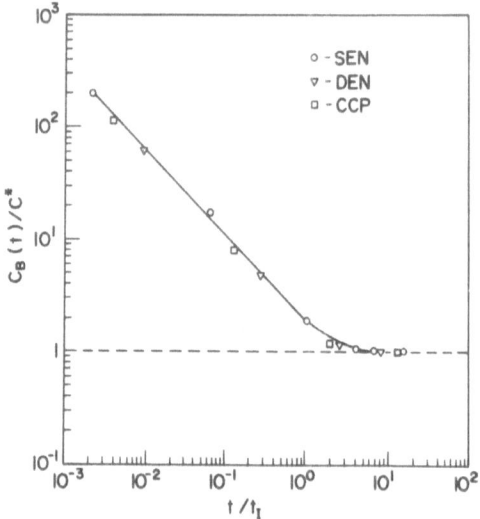

Fig. 3:   When C_B is normalized by C*, and t by t_I, all geometries yield a single curve.

3. CREEP CRACK GROWTH

The propagation of cracks during creep in structural materials occurs by the nucleation, growth and coalescence of cavities on grain boundaries. This process has received considerable theoretical attention, from both mechanistic[5-8] and continuum mechanical[9] viewpoints. The work on micromechanisms, which deals with cavitation under uniform stresses, shows that the process may be stochastic in nature, and is not dependent on a single component of the stress tensor. There is still some debate over the correlation between these models and experimental results for well-characterized materials, tested using uniform stress states. It is not surprising therefore, that the incorporation of these models into a model for creep crack growth is at an early stage. Those models which have been developed assume a steady-state stress field is established ahead of the crack tip.[10,11]

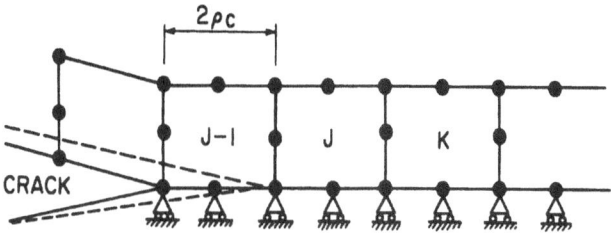

Fig. 4: An illustration of the procedure by which cracks are allowed to propagate.

An alternative approach involves the use of continuum parameters for damage. This can more readily be incorporated into a finite element model for creep crack growth, and therefore offers the possibility of studying crack propagation in a time-dependent stress field. We have adopted this approach in the current study.[12] The general methodology is illustrated in Fig. (4). A finite element mesh is created which contains a row of 8 node isoparametric elements, each of length ℓ, ahead of the crack, in the crack plane. The crack is assumed to propagate in steps of ℓ, whenever the damage within the element just ahead of the crack tip reaches a critical value D_c. We define D_{kj} and $\dot{D}_{kj}$ as the current amount of damage and its rate of accumulation in element k, when the current tip is located at element (j-1). If the time interval between the failure of element (j − 1) and that of element j is t_j, then the average crack growth rate is equal to $(da/dt) = \ell/t_j$. In the model, the critical level of damage D_c, is assumed to be a material parameter, which needs to be determined experimentally. The characteristic distance ℓ, is also a material parameter, although the choice of ℓ is limited in part by practical limitations on the smallest mesh size which is economically viable.

In the present study we have used the method to assess the creep crack growth behaviour of coarse-grained $2\frac{1}{4}$ Cr-1 Mo steel (simulated heat-affected zone material). The damage parameter we have used is the von Mises equivalent tensile strain ε_e. The justification for this is based on experimental data from smooth tensile and plane strain creep tests, at 550°C (for details, see ref. 12). The amount of accumulated damage (defined either by the number of cavities per unit area or the cavitated area fraction) correlates poorly with the principal tensile strain. However, as Fig. (5) indicates, good correlation is obtained with the equivalent tensile strain. These results are in agreement with those of others,[13-15] including tests done in torsion. From Fig. (5) it appears that damage accumulates slowly up to an effective strain of between 2 and 3%, whereupon an acceleration of damage commences.

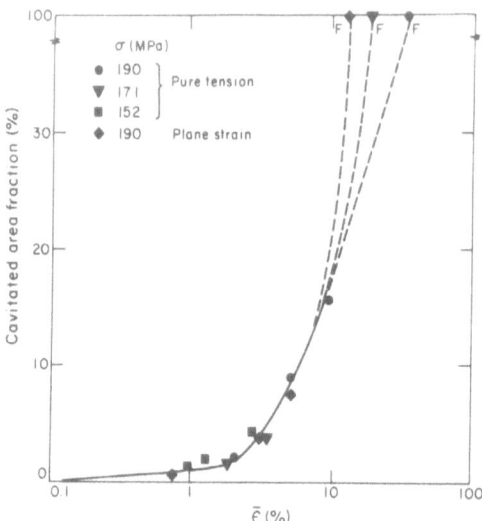

Fig. 5:   When damage (in this case the cavitated area fraction) is plotted against effective strain, a single curve is obtained.

The final failure strain is about 20%. However, for material ahead of a crack tip, the load bearing capacity will have disappeared long before this strain is approached. Moreover, the rate of damage at large strains is governed by tertiary creep, which the finite element analysis does not incorporate. We therefore define a critical strain for material exhaustion ahead of the crack tip. In the present work, two values for this parameter are used - 1% and 5%. The second of these was selected in all our early work, as representing the onset of failure in the coarse-grained $2\frac{1}{4}$ Cr - 1 Mo steel studied experimentally. The size of the finite element mesh on the crack plane was equal to 0.5 mm, or about two grain diameters.

The propagation of creep cracks using these conditions has been investigated using the finite element method. Details of this work are contained in ref. (16). Two specimen geometries - namely SEN and DEN, were investigated. In both cases, cracks did not begin to propagate until after establishment of a steady-state stress field. The solutions obtained are compared with experimentally determined crack growth measurements available in the literature, in Figs. (6) to (8). There are several points to note. The three figures represent material in different heat treatment conditions. In particular the work of Siverns and Price[17] (Fig. 6) utilized a lightly tempered heat of $2\frac{1}{4}$ Cr-1 Mo steel (1 hour at 690°C). The expected ductility of this material is very small, and a critical strain of 5% (used in the FEM solutions) is no doubt too high. This explains the discrepancy between the FEM solutions and experimental results. For the other two sets of data, Figs. (7) and (8) the material was in a heat treatment condition close to that used in our creep tests. In this case, the agreement with the FEM solutions is excellent. Figs. (7) and (8) also illustrate that the FEM results are independent of geometry when plotted against the steady-state C*.

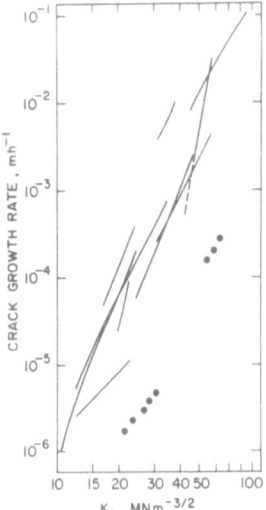

Fig. 6: Comparison of the numerical results (closed circles), with the experimental data of Siverns and Price,[17] for a brittle heat of $2\frac{1}{4}$ Cr- 1 Mo steel.

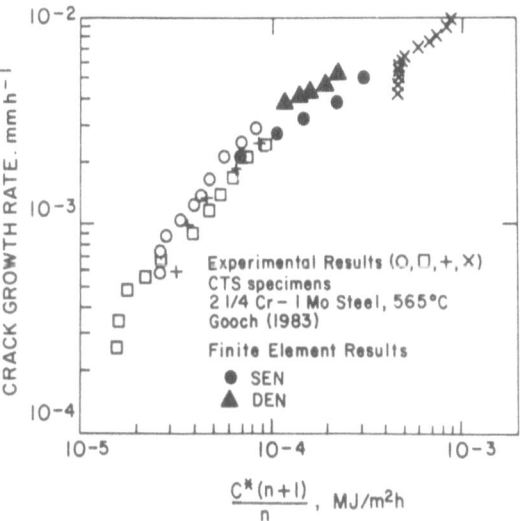

Fig. 7: Comparison of the numerical results with the experimental results of Gooch.[18]

8

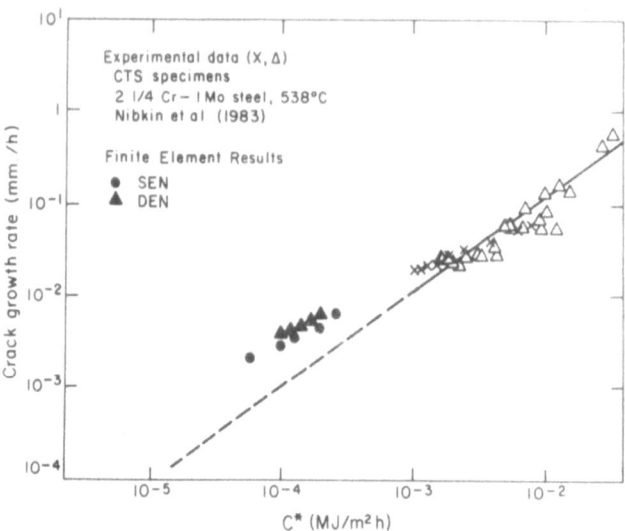

Experimental data (X, Δ)
CTS specimens
2 1/4 Cr − 1 Mo steel, 538°C
Nibkin et al (1983)

Finite Element Results
● SEN
▲ DEN

Fig. 8:   Comparison of the numerical results with the experimental results of Nibkin et al.[19]

These results demonstrate the potential for assessing the growth of cracks based on a careful analysis of damage accumulation in small specimens. The procedure could possibly be used also for assessing the structural integrity of service-exposed material, based on accelerated creep tests, and a knowledge of the current state of damage.

4. CREEP CRACK GROWTH UNDER TRANSIENT CONDITIONS

We now turn our attention to the correlation of creep crack growth, under conditions where a steady-state field is not established prior to propagation. To study this, we decrease the critical equivalent strain for material failure to 1%. The results, using the same procedure as outlined in the previous section are shown in Fig. (9). Initially, the crack growth rate decreases due to gradual stress relaxation as steady-state is approached. As the crack length increases however, the net section stress on the remaining ligament rises, and the crack starts to accelerate.

Figure (10) shows the variation of the C-integral, calculated over several different paths, as a function of crack length, for the DEN geometry. The C-integral eventually becomes path independent, but only after considerable crack growth. For the SEN geometry, some path dependence occurs throughout crack growth. In both cases, the near-field integral (i.e. C_B) is initially higher than the far-field integrals, but drops below them before path independent behaviour is established.

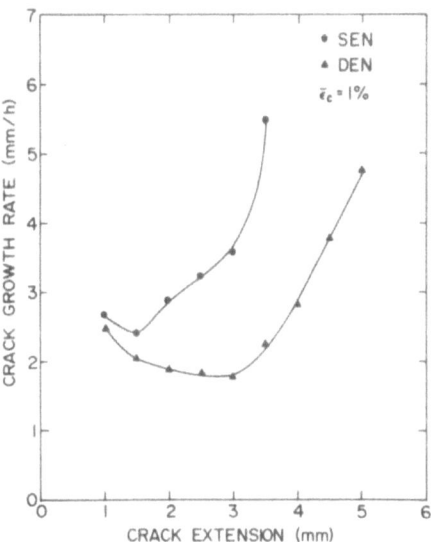

Fig. 9: The variation of crack growth rate with crack extension for both the SEN and DEN geometries.

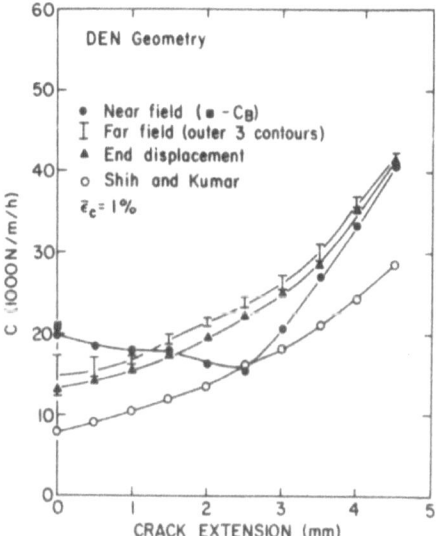

Fig. 10: The variation of the C-integral with crack extension for different paths ranging from near field (C_B and the inner contour) to far field (outer most contour).

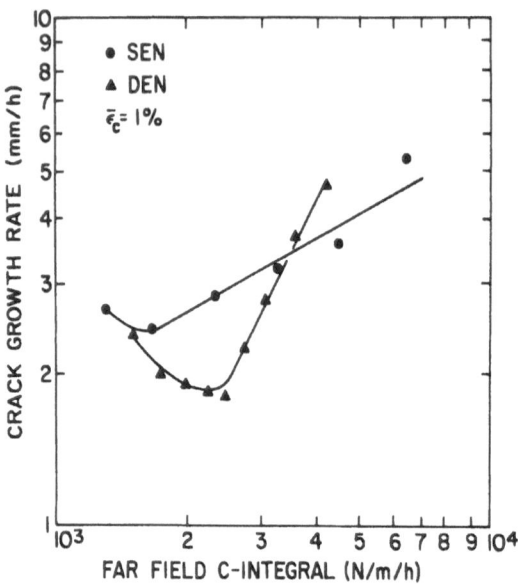

Fig. 11: No correlation is found when crack growth rate is plotted against the far-field C-integral.

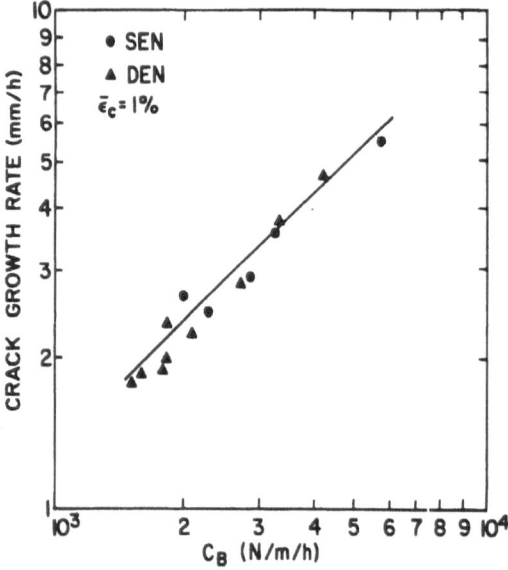

Fig. 12: Good correlation is found between crack growth rate and the C_B integral.

We have attempted to correlate the crack growth rates determined for both geometries with a variety of parameters (K_I, far-field C-integral, crack opening displacement rate, and C_B). Full details are presented in ref. (16). Of these four parameters, only C_B provides any degree of correlation. Figure (11) is typical of the behaviour of the other three parameters. It shows the crack growth rate, plotted against the far-field C-integral for the two geometries. This is essentially the value of C obtained experimentally using the load-point displacement rate. It is clear that no correlation is possible. All three of these parameters rise monotonically during crack growth, despite the initial decrease in crack growth rate, and thus give u-shaped curves. The C_B (i.e. near-field) integral, on the other hand, does correlate the crack growth rates (Fig. (12)). We therefore conclude that while none of the conventional parameters is capable of correlating creep crack growth under non-steady stress-strain fields, a parameter based on the near-crack tip field does have this capability. It must be noted however, that the value of C_B during crack growth is not equal to that determined after the same time for a stationary crack. It is not possible therefore, to use Fig. (3) to estimate the value of C_B as a function of time.

5. CONCLUSIONS

a)   The transient stress-strain field ahead of a stationary crack in a creeping solid is well described by the HRR field, with an amplitude equal to C_B. C_B can be estimated as a function of K_I and the steady-state C^*, as shown in Fig. (3).

b)   A continuous damage parameter, based on microstructural observations from tensile creep tests, can be used in conjunction with the finite element method, to describe creep crack growth under steady-state conditions. Good correlation is obtained with experimental creep crack growth data.

c)   When cracks grow before a steady-state field is established, none of the usual parameters (K_I, COD rate, or far-field C-integral) adequately correlate with the crack growth rates. The near-field C-integral, which we call C_B, does provide adequate correlation.

REFERENCES

1.   H. Reidel and J.R. Rice, ASTM STP 700 (1980), 112.

2.   S.B. Biner, D.S. Wilkinson and D. Watt, Proc. CFC 10 (1984), in press.

3.   V. Kumar and C.F. Shih, EPRI Report No. 1931/1237-1 (1981).

4.   R.B. Stonesifer and S.N. Atluri, Engg. Fract. Mech., 16 (1982), 625.

5.   A.S. Argon, I.-W. Chen and C.-W. Lau, Creep-Fatigue-Environment Interactions (AIME, 1980), 46.

6.   A. Needleman and J.R. Rice, Acta Met., 28 (1980), 1375.

7.   A.C.F. Cocks and M.F. Ashby, Prog. Mater. Sci., 27 (1982), 189.

8.   T. Takesugi and V. Vitek, Met. Trans. **12A** (1981), 659.

9.   L.M. Kachanov in "Problems of Continuum Mechanics", (Edit. J.RM. Radok).

10. V. Vitek, Acta Met., **26** (1978), 1345.

11. D.S. Wilkinson and V. Vitek, Acta Met., **30** (1982), 1723.

12. S.B. Biner and D.S. Wilkinson, to be published.

13. B.J. Cane, Proc. ICF-5 (Cannes, 1980), 1285.

14. B.F. Dyson, Metal Sci. J., **10** (1976), 349.

15. D. McLean, B.F. Dyson and D.M.R. Taplin, ICF-4 (Waterloo, 1977), 325.

16. S.B. Biner and D.S. Wilkinson, to be published.

17. M.J. Siverns and A.T. Price, Intl. J. Fract., **9** (1973), 199.

18. D.J. Gooch and B.L. King, CERL Report RD/L/R/1976.

19. K.M. Nibkin, D.J. Smith and G.A. Webster, Proc. ASME, Intl. Conf. on "Advances in Life Prediction Methods", (ed. D.A. Woodford and J.R. Whitehead, 1983), 249.

ACKNOWLEDGEMENTS

This work was supported by NSERC, Canada and by the Department of Energy, Mines, and Resources (through DSS Contract 05SU.23440-0-9199).

THE CYCLIC DEFORMATION OF 70-30 α-BRASS: INTERNAL AND EFFECTIVE STRESSES AND DISLOCATION SUBSTRUCTURES

J. BOUTIN*, J.I. DICKSON and J.-P. BAÏLON

ECOLE POLYTECHNIQUE, Montréal, Québec, Canada, H3C 3A7
*Present address: PMRL, Canmet, Ottawa, Ontario, Canada

ABSTRACT

The cyclic deformation at controlled strain amplitudes, $\Delta\varepsilon/2$, of 70-30 α-brass was studied employing measurements of cyclic internal and effective stresses, based on the detection of reversed microplasticity on the hysteresis loops. The effective stress increases slowly with the number of cycles, is independent of grain size and near the end of the fatigue life is almost independent of $\Delta\varepsilon/2$. Its behaviour indicates that it is controlled by the interaction of dislocations with solute atoms and with debris produced during cycling. The internal stress appears controlled by the dislocation substructure. Its variation is responsible for a maximum in the cyclic stress vs log N curves at high $\Delta\varepsilon/2$ and for an intermediate plateau in the cyclic stress-strain curve. Both effects result from increased recovery during cycling. The influence of the precision of the hysteresis loop records on the results is examined. The dislocation substructures observed, consisting largely of edge dislocations and indicating the occurrence of some recovery, are related to the cyclic behaviour.

1. INTRODUCTION

One of the time-dependent phenomena that can influence fatigue mechanisms is thermally activated cyclic deformation. To study this phenomenon it is important to separate the cyclic stress into an internal and an effective stress component. As summarized elsewhere (1), there has been considerable recent interest in decomposing the cyclic stress into these components (e.g., 1-8) or into related friction and back stresses (e.g., 9, 10). The techniques that have been employed to measure cyclic internal and effective stresses include those previously employed in monotonic deformation studies, such as stress relaxation (5), partial unloading to zero stress relaxation (4) and strain rate changes (6). Such techniques are tedious and often experimentally difficult to apply to cyclic deformation. For these reasons, simple techniques based on the analysis of the shape of the hysteresis loops are particularly attractive. Two such techniques have been proposed. That employed by Kuhlmann Wilsdorf and Laird (9) and based on Cottrell's analysis (11) of the hysteresis loop measures a friction and a back stress. The technique developed by Handfield and Dickson (7), which we shall refer to as the HD technique, attempts to approximate a partial unloading to zero stress relaxation and consists (1, 7) in determining the mid-point of the most elastic segment of the hysteresis loop and taking the effective stress, $\Delta\sigma^*/2$, as the difference between the peak stress and this mid-point stress and the internal stress $\Delta\sigma_i/2$ as the difference between this mid-point stress and the mean stress. The similarities and differences between the internal and back stresses and the effective and friction stresses measured by these two simple techniques have been recently discussed (1). The precise measurement of these stress components by the HD technique will be particularly difficult when the thermal stress component is small. As an example, the relative values but not the variation during a test, of the internal and effective stresses for polycrystalline copper deformed at 20-22°C were found to be very sensitive to the experimental accuracy employed (1).

The primary objectives of the present study are to demonstrate the information obtainable by applying the HD technique to the room temperature cyclic deformation of 70-30 α-brass and to relate these results to the dislocation substructures observed by transmission electron microscopy (TEM). A secondary objective is to evaluate the influence of the precision of the hysteresis loop records on the results.

2. EXPERIMENTAL PROCEDURE

The 70-30 α-brass employed is the same as described by Marchand et al (12). The experimental procedure employed is also similar and is only briefly summarized. The material was given combinations of mechanical (cold rolling) and thermal treatments to obtain average grain sizes, d, of 45, 120 and 260 μm.

The cylindrical specimens were tested in fully reversed total strain control on an Instron servohydraulic machine interfaced to a microprocessor for test control and data acquisition. A symmetrical saw-tooth waveform with a constant total strain rate of rate of 0.01 s^{-1} was employed. For the 45 and 260 μm grain sizes, 60 pairs of (σ, ε) data points were digitally recorded per hysteresis loop. Subsequently, to improve the precision for the tests on the 120 μm grain size, 200 pairs of data points per measured hysteresis loop were recorded. As well, to obtain a precise value of the experimental Young's modulus, each test was preceded by recording 10 hysteresis loops at a total strain amplitude, $\Delta\varepsilon/2 = 0.02\%$.

3. RESULTS

3.1 Cyclic Stress Behaviour

The curves of the cyclic stress $\Delta\sigma/2$ versus log number of cycles N are presented in Figures 1 and 2 for d = 45 and 260 μm, respectively. Three types of curves can be seen similar to the results of Marchand et al (12). For low amplitudes ($\Delta\varepsilon/2 < 0.25\%$) the cyclic stress increases throughout the test and shows little tendency to saturate. For intermediate amplitudes ($0.25\% < \Delta\varepsilon/2 < 0.40\%$), rapid cyclic hardening occurs initially followed by apparent saturation starting at about 100 cycles. For high amplitudes ($\Delta\varepsilon/2 > 0.4\%$), $\Delta\sigma/2$ shows initial rapid hardening before passing through a maximum followed by cyclic softening and saturation or slow rehardening.

Typical curves of the cyclic internal and effective stresses evaluated by the HD technique are plotted against log N for these two grain sizes and are presented in Figures 3 and 4. For the 120 μm grain size specimens, the more precise hysteresis loop records and experimental Young's modulus measurements permitted more accurate determination of the start of reversed microplasticity. The values of $\Delta\sigma_i/2$ and $\Delta\sigma^*/2$ obtained on these samples are shown in Figure 5. At all amplitudes, $\Delta\sigma^*/2$ slowly increases with increasing strain amplitude and does not tend to saturate prior to fracture. At low amplitudes, $\Delta\sigma_i/2$ either remains approximately constant or initially increases and then saturates. At higher amplitude, $\Delta\sigma_i/2$ tends to pass through a maximum. The more precise results for the 120 μm grain size show lower values of and a smaller increase in $\Delta\sigma^*/2$ through a cyclic test. If the effective stress just prior to failure is compared for the different tests performed, these higher precision results indicate that this value of $\Delta\sigma^*/2$ is almost independent of $\Delta\varepsilon/2$. While $\Delta\sigma^*/2$ varies more rapidly at the higher amplitudes, this effect is offset by the shorter lifes. The values of $\Delta\sigma^*/2$ at the start of cycling increase somewhat with amplitude for $\Delta\varepsilon/2 > 0.25\%$.

Figure 6 presents the cyclic stress-strain (CSS) curves with the plastic strain amplitude, $\Delta\varepsilon_p/2$, taken as the half-width of the hysteresis loops at the mean stress. The CSS curves present intermediate plateaus for each grain size, similarly to that previously observed by Marchand et al (12) for d = 120 μm. Figure 7 presents the corresponding CSS curves of $\Delta\sigma^*/2$ and $\Delta\sigma_i/2$ vs log $\Delta\varepsilon_p/2$ for d = 45 and 260 μm. The grain size strongly influences the internal stess except for the highest strain amplitudes employed

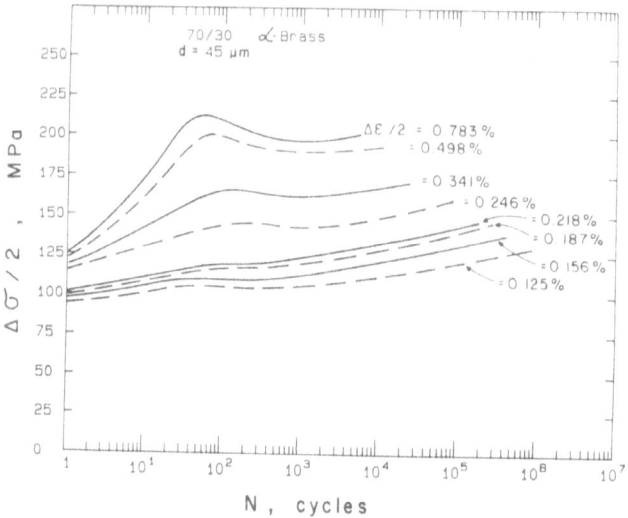

Fig. 1 - Curves of $\Delta\sigma/2$ vs log N, d = 45 μm.

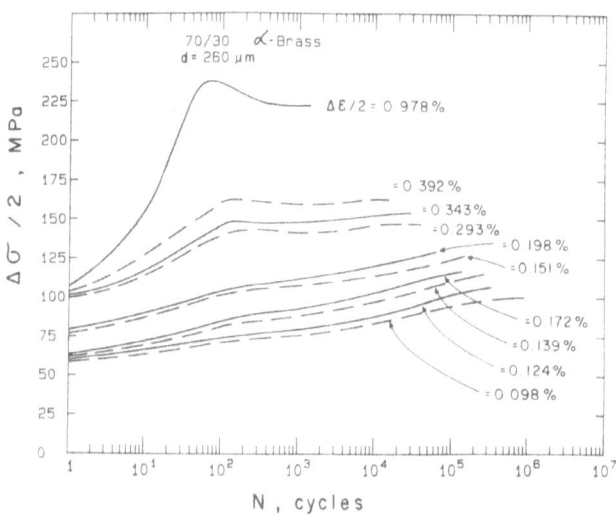

Fig. 2 - Curves of $\Delta\sigma/2$ vs log N, d = 260 μm.

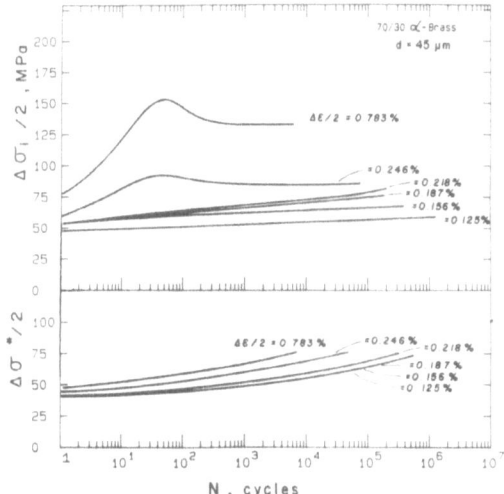

Fig. 3 - Typical curves of $\Delta\sigma_i/2$ and
$\Delta\sigma^*/2$ vs log N, d = 45 μm.

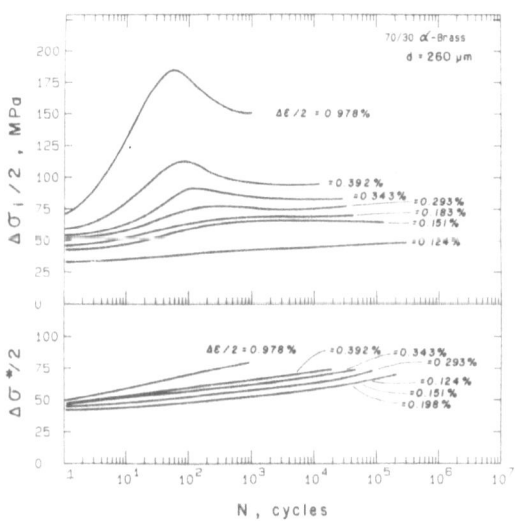

Fig. 4 - Typical curves of $\Delta\sigma_i/2$ and
$\Delta\sigma^*/2$ vs log N, d = 260 μm.

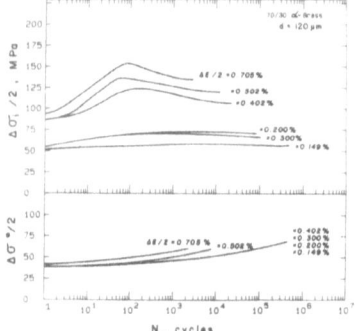

Fig. 5 Curves of $\Delta\sigma_i/2$ and $\Delta\sigma^*/2$ vs log N, d = 120 μm.

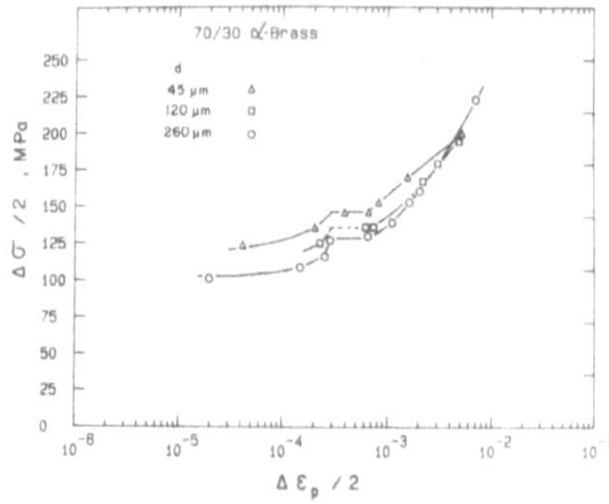

Fig. 6 CSS curves for the three grain sizes.

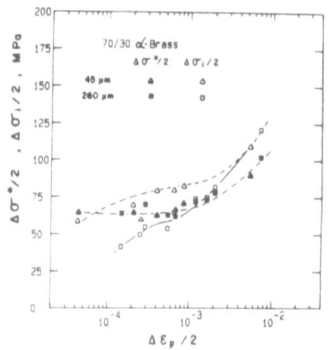

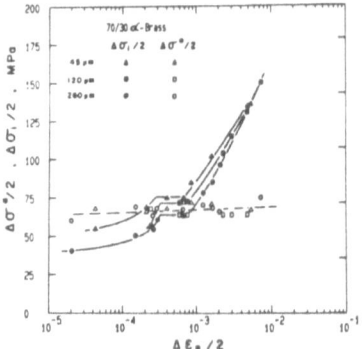

Fig. 7 CSS curves of $\Delta\sigma_i/2$ and $\Delta\sigma^*/2$ vs $\Delta\varepsilon_p/2$ for d = 45 and 260 μm.

Fig. 8 CSS curves of $\Delta\sigma_i/2$ and $\Delta\sigma^*/2$ for d = 120 μm and retreated d = 45 and 260 μm data.

but has little influence on $\Delta\sigma^*/2$.

To improve the precision of the $\Delta\sigma_i/2$ and $\Delta\sigma^*/2$ measurements for the 45 and 260 μm grain size materials, the hysteresis loops were retreated for different Young's moduli within the range (107± 1 GPa) measured for the 120 μm grain size. The correct modulus was taken as that which gave the longest elastic segment of hysteresis loop. The results obtained based on a more severe criterion for the start of reversed microplasticity for d = 45 and 260μm, are compared to the more precise results for 120 μm grain size material on a CSS basis in Figure 8. These results confirm the lack of influence of grain size on the near fatigue fracture value of $\Delta\sigma^*/2$ and also indicate a striking lack of influence of strain amplitude on this $\Delta\sigma^*/2$ value.

3.2 Microscopic Observations

The slip line patterns observed by optical microscopy showed that for $\Delta\epsilon/2$ values below the plateau on the CSS curve, dark, pronounced slip bands were only present within isolated individual grains. These slip bands were shown by scanning election microscopy to contain intrusion-extrusion pairs similar to those observed in copper (13). The CSS plateau corresponded to more macroscopic bands a few grains in width, in which most grains contained but were not completely covered by such bands, separated by regions in which few grains contained these slip bands. As well, quite frequently such pronounced slip bands in neighbouring grains corresponded at grain boundaries. For $\Delta\epsilon/2$ values equal or greater than the plateau amplitudes, cross-slip traces were commonly observed and double slip became more apparent with increasing amplitude. As $\Delta\epsilon/2$ increased on and past the plateau, the regions in which most grains contained pronounced slip bands covered an increasing percentage of the specimen surfaces.

The dislocation substructures produced in the different regions of the CSS curve were studied in thin foils taken perpendicularly to the stress axis. For amplitudes below the plateau, the general feature was that of very planar slip, with parallel dislocations (Figure 9), which appeared edge in character, present in individual parallel slip bands. In regions where the dislocation density in these bands was lower, stacking faults were often seen separating partial dislocations. On the plateau region, the observed dislocation densities were higher. Regions were observed with dislocations of two orientations consistent with edge dislocations left behind by cross-slipping screw dislocations (Figure 10). As well, observations were obtained of slip bands in which recovery had occurred in neighbouring grains coinciding at grain boundaries (Figure 11). Above the plateau, the dislocation density increased (Figure 12) and there was a tendency for parallel

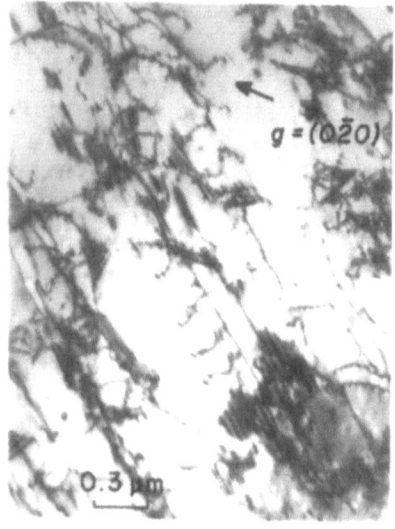

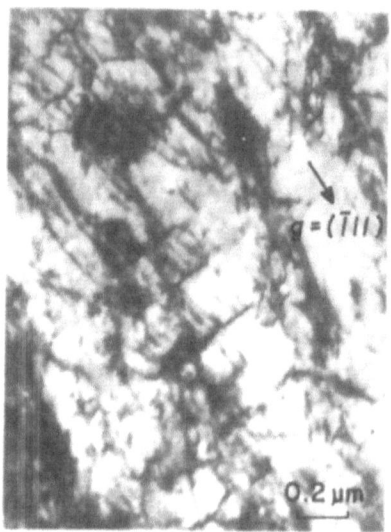

Fig.9 Arrangements of dislocations and stacking faults in slip
bands, $\Delta\varepsilon/2$ = 0.183% d = 260 μm.

Fig.10 Two dominant dislocation orientations suggest edge
dislocations left by cross-
slip, $\Delta\varepsilon/2$ = 0.246%, d=45 μm.

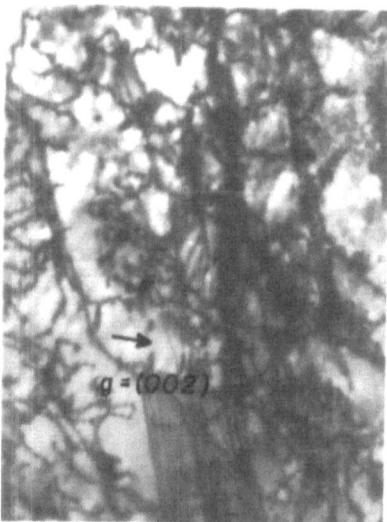

Fig. 11 Spread of pronounced
slip bands across grain bound-
aries, $\Delta\varepsilon/2$ = 0.246%, d = 45μm.

Fig. 12 Dislocation arrangement suggesting start of rough
polygonization, $\Delta\varepsilon/2$ = 0.343%,
d = 260 μm.

dislocations to be arranged in bundles suggesting rough polygonization. Again most dislocations appeared edge in character.

4. DISCUSSION

4.1 Internal and Effective Stresses

The $\Delta\sigma_i/2$ and $\Delta\sigma^*/2$ measurements for 70-30 α-brass cycled at
$20-22^0C$ indicate that the dislocation substructures strongly in-
fluences the internal stress which increases strongly with strain
amplitude and is grain size dependent. The variation of internal
stress is responsible for the intermediate plateau in the CSS
curve and for the maximum in the $\Delta\sigma/2$ vs log N curve for the higher $\Delta\varepsilon/2$ values. The effective stress is more weakly dependent on
$\Delta\varepsilon/2$, apparently independent of grain size and generally does not
follow the variations in $\Delta\sigma_i/2$. These results are consistent with
the hypothesis that $\Delta\sigma^*/2$ in 70-30 α-brass is largely determined
by dislocation - Zn solute atom interactions in agreement with the
identification (14) of the rate-controlling mechanism for the
monotonic deformation kinetics of this material for temperatures
< 100^0C. The result that $\Delta\sigma^*/2$ increases with log N but generally
not in the same manner as $\Delta\sigma_i/2$ suggests that an additional significant contribution to $\Delta\sigma^*/2$ is the interaction between the mobile
dislocations and the dislocation debris produced by the cyclic
deformation. The indication is that $\Delta\sigma^*/2$ is determined by the
interaction between the mobile dislocations and the combination of
these two types of defects (solute atoms and dislocation debris).

4.2 The Occurrence or Absence of Stress Saturation

Marchand et al (12) have questioned whether saturation of the
cyclic stress at low $\Delta\varepsilon/2$ values in 70-30 α-brass truly occurs.
The present results confirm this absence of saturation at low
amplitudes. The measured variations of $\Delta\sigma^*/2$ vs log N in fact
indicate that true cyclic stress saturation was not obtained prior
to fracture at any $\Delta\varepsilon/2$ employed. Apparent saturation of $\Delta\sigma/2$
occurs at intermediate amplitudes since the cyclic softening past
the slight maximum in the $\Delta\sigma_i/2$ vs log N curve is offset by the
$\Delta\sigma^*/2$ hardening. This apparent saturation of $\Delta\sigma/2$ occurs approximately at 0.1% of the fatigue compared to 1% for copper at similar
(saturated) $\Delta\varepsilon_p/2$ values. Even through the fatigue life was approximately ten times longer for the 70-30 α-brass than for copper, the indication remains that saturation is occurring much too
rapidly for this low stacking fault energy (SFE) material.

4.3 The Dislocation Substructures

The dislocation substructures observed depending on amplitude
show similarities with those reported by Youssef (15) and Lukas
and Klesnil (16) for 70-30 α-brass single crystals. At low ampli-

tudes, these substructures typify the occurrence of limited amounts of very planar cyclic slip. The dislocation substructures observed in the intermediate plateau confirm the optical microscopy observations. A considerable amount of cross slip occurs in the intermediate plateau and this results in the occurrence of some recovery. This recovery is particularly important in some slip bands in which the cyclic deformation can then become more concentrated. This results in the formation of extrusion-intrusion pairs at the sample surface making these bands easily distinguishable by optical microscopy. The TEM observations at grain boundaries suggests that recovery which has occurred in slip bands on one side of the boundary can provoke recovery in slip bands on the opposite side probably because the slip remains planar resulting in dislocation pile-ups at the grain boundary. The high dislocation densities observed above the plateau in the CSS curve are in agreement with the strong cyclic hardening obtained. The dislocation substructures observed also suggest that some polygonization and a significant amount of recovery has occurred which, however, is much less complete than that obtained in higher SFE materials such as copper.

The other important result, which has been confirmed by L'Espérance's (17) more detailed analysis of the dislocation substructures observed and also agrees with the observations of Lukas and Klesnil (16), is that the dislocations observed are generally edge in character, indicating that the screw dislocations are more mobile. This result supports the proposal that one of the major contributions to $\Delta\sigma^*/2$ is the interaction between the mobile dislocations and the solute atoms. Although the dislocations are assumed separated into Schockley partials whose characters differ by + and -30^0 from that of the a/2 <110> dislocations, the indication is that the interaction with the solute atoms remains stronger with the dissociated edge dislocations.

4.4 The maximum in the $\Delta\sigma/2$ vs log N curves

A maximum in the $\Delta\sigma/2$ vs log N curves of α-brass was obtained for $\Delta\varepsilon/2 > 0.40\%$, in agreement with the previous results of Marchand et al (12). Its occurrence is associated with a clearer maximum in the $\Delta\sigma_i/2$ vs log N curves, which can be observed for $\Delta\varepsilon/2 > 0.25\%$, i.e., for all values above the plateau on the CSS curve. Similar maxima in $\Delta\sigma/2$ have been observed for 304 (18) and 316 (19) stainless steels and for commercial-purity titanium (7,8). In the 316 stainless steel, this maximum was also shown to result from a maximum in the $\Delta\sigma_i/2$ curve. In Ti, the cyclic softening after the maximum was shown (7,8) to correspond to an increase in the mobility of screw dislocations and in their resulting mutual annihilation by cross-slip. In all these examples, the recovery mechanisms operate with difficulty especially at the start of cycling which results in very rapid cyclic hardening at suffi-

ciently high amplitudes. Continued strain cycling, however, acti-
vates increased recovery. In the low SFE f.c.c. metals, this
results in a significant decrease in $\Delta\sigma_i/2$. In the case of titanium as well as of zirconium (7,8) cyclic softening corresponded
to a decrease in $\Delta\sigma^*/2$, which result has considerable implications
(8) concerning the mechanism controlling the deformation kinetics
of these two h.c.p. metals.

4.5 The Intermediate Plateau

The intermediate plateau in the curves of $\Delta\sigma/2$ and $\Delta\sigma_i/2$ vs
$\Delta\varepsilon_p$ also only appears to have been previously reported for α-brass
by Marchand et al (12) who clearly observed it on the CSS curve
for 120 μm grain size material. Its characteristics appear very
similar to those of the intermediate plateau observed by Sinning
and Haasen (20) on solution-treated Cu-4% Ti of 110 μm grain size.
It is essentially associated (20, 21) with the spread of pronounced slip bands, in which significant recovery has occurred, across
grain boundaries. This local recovery in these slip bands is
first activated at grain boundaries in these low SFE materials.
Its occurrence permits the cyclic slip to concentrate itself in
these bands, resulting in the formation of intrusion-extrusion
pairs at the surface. In this aspect, the pronounced slip bands
are similar to classical persistent slip bands found in high SFE
materials, however, the recovery in the bands observed in 70-30 α-
brass did not give rise to the ladder-like substructure.

4.6 The usefulness of the HD technique

The present study shows that the cyclic internal and effective stresses measured employing the HD technique permit the obtention of important information concerning the cyclic behaviour
of 70-30 α-brass which is consistent with the dislocation substructures observed and with the literature results on the monotonic deformation behaviour of this material. This technique has
now been employed to study cyclic deformation for five different
polycrystalline materials. For 316 stainless steel, a maximum in
$\Delta\sigma/2$ vs log N curves was also shown to correspond to a maximum in
the $\Delta\sigma_i/2$ vs log N curves. For polycrystalline copper, the $\Delta\sigma^*/\Delta\sigma$
ratio was shown to remain almost constant in keeping with this
material obeying a Cottrell-Stokes law (22). For commercial purity Ti and Zr (7,8,) the anomalous cyclic softening behaviour was
shown to correspond to a decrease of $\Delta\sigma^*/2$ which, combined with
the changes in dislocation substructures observed, gave new insight into the thermally activated deformation of these two materials. These studies have demonstrated different $\Delta\sigma^*/2$ and
$\Delta\sigma_i/2$ behaviour for the different types of material which in every
case was consistent with other results or observations.

The difference between the ordinary and high precision results on α-brass, similar to previous studies on Cu (1) and Ti (23), is that the ordinary precision recordings of hysteresis loops allow to follow reasonably well the variations of $\Delta\sigma^*/2$ and $\Delta\sigma_i/2$ with log N or with $\Delta\varepsilon/2$. High precision recordings however are important in obtaining reliable measurements of the relative values of $\Delta\sigma_i/2$ and $\Delta\sigma^*/2$, as well as in studies where higher precision is required in studying the variations of $\Delta\sigma^*/2$ or $\Delta\sigma_i/2$ as a function of a given parameter.

The HD technique is also closely related (1) to that employed by Kuhlmann-Wilsdorf and Laird (9) in the case of a small thermal stress component provided that the same criterion for the start of reversed plasticity is employed. If it is accepted that a partial unloading to zero stress relaxation method is a reliable method of measuring internal and effective stresses, and there is evidence to this effect (4, 8, 24), then the measurements should be based on the start of reversed microplasticity. Recent measurements on 316 stainless steel (25) have shown that the HD technique employed on high precision recordings of hysteresis loops gives very good agreement with $\Delta\sigma_i/2$ and $\Delta\sigma^*/2$ values determined by partial unloading to zero stress relaxation. The most probable explanation (1, 4) of the significantly lower $\Delta\sigma_i/2$ values measured with the technique employed by Kuhlmann-Wilsdorf and Laird (9), where there is a tendency to base the results on the start of reversed macroplasticity, is that rapid relaxation of the internal or back stress occurs between the start of reversed microplasticity and that of reversed macroplasticity. To obtain reliable measurements of $\Delta\sigma^*/2$ and $\Delta\sigma_i/2$ based on the start of reversed microplasticity requires high precision recordings of hysteresis loops and high precision measurements of the (experimental) elastic modulus. The influence on the values of $\Delta\sigma^*/2$ can be seen in comparing Figures 7 and 8 and has also been demonstrated previously for polycrystalline copper (1).

5.   CONCLUSIONS

From the present study, it can be concluded that the measurement of cyclic internal and effective stress for α-brass gave reliable and useful information on the cyclic deformation behaviour, which could also be related to the dislocation substructures observed.

6.   ACKNOWLEDGMENT

The authors wish to thank Gilles L'Espérance for his assistance with the TEM observations. Financial support from the FCAC (Quebec) and NSERC (Canada) programs is gratefully acknowledged. Gratitude is expressed to the Association des Diplômés of Ecole Polytechnique who awarded Jacques Boutin a graduate scholarship.

REFERENCES

1. Dickson, J.I., Boutin, J. and Handfield, L. A Comparison of Two Simple Methods for Measuring Cyclic Internal and Effective Stresses. Materials Science and Engineering 64 (1984) L7-L11.
2. Polak, J. and Klesnil, M. The Hysteresis Loop 1. A Statistical Theory Fatigue of Engineering Materials and Structures 5 (1982) 19-32.
3. Polak, J., Klesnil, M. and Helesic, J. The Hysteresis Loop 2. An Analysis of the Loop Shape. Fatigue of Engineering Materials and Structures 5 (1982) 33-44.
4. Polak, J., Klesnil, M. and Helesic, J., The Hysteresis Loop 3. Stress-Dip Experiments. Fatigue of Engineering Materials and Structures 5 (1982) 45-56.
5. Tsou, J.C. and Quesnel, D.J. Internal Stress Measurements During the Saturation Fatigue of Polycrystalline Aluminum, Materials Science and Engineering 56 (1982) 289-299.
6. Miura, S. and Umeda, K. Variation of Effective and Internal Stresses During Cyclic Straining in α-Zirconium. Scripta Metallurgica 7 (1973) 337-344.
7. Handfield, L. and Dickson, J.I. The Cyclic Deformation of Titanium: Dislocation Substructures and Effective and Internal Stresses. Defects, Fracture and Fatigue. (The Hague, Martinus Nijhoff, 1983), pp 37-51.
8. Dickson, J.I., Handfield, L. and L'Espérance, G. Cyclic Softening and Thermally Activated Deformation of Zr and Ti. Materials Science and Engineering 60 (1983) L3-L7.
9. Kuhlmann-Wilsdorf, D. and Laird, C. Dislocation Behaviour in Fatigue II. Friction and Back Stress as Inferred from an Analysis of Hysteresis Loops. Materials Science and Engineering 37 (1979) 111-120.
10. Cheng, A.X. and Laird, C. Mechanisms of Fatigue Hardening in Copper Single Crystals: The Effects of Strain Amplitude and Orientation. Materials Science and Engineering 51 (1981) 111-121.
11. Cottrell, A.H. Dislocations and Plastic Flow in Crystals (London, Oxford University Press, 1953, pp 111-132).
12. Marchand, N., Baïlon, J.-P. and Dickson, J.I. The Cyclic Response and Strain Life Behaviour of Polycrystalline Copper and α-Brass. Defects, Fracture and Fatigue. (The Hague, Martinus Nijhoff, 1983) pp 195-208.
13. Hunsche, A. and Neumann, P. On the Formation of Extrusion-Intrusion Pairs During Fatigue of Copper. Advances in Fracture Research, Vol. 1 (Oxford, Pergamon Press, 1981) 273-279.
14. Flor, H. and Neuhauser, H. Quantitative Analysis of Stress Relaxation in Cu-30% Zn Single Crystals: Variation of Obstacle Strength and Athermal Stress. Acta Metallurgica 28 (1980) 939-948.
15. Youssef, T.H. Dislocations in α-Brass Single Crystals. Physica Status Solidi (a) 3 (1970) 801-810.

16. Lukas, P. and Klesnil, M.   Dislocation Structures in Cu-Zn. Physica Status Solidi 37 (1970) 833-842.
17. L'Espérance, G.   Ecole Polytechnique de Montréal, unpublished research.
18. Turner, A.P.L., Cyclic Deformation Behavior of Type 304 Stainless Steel at Elevated Temperature.   Metallurgical Transactions 10A (1979) 225-234.
19. Bernard, M., Vogt, J.B., Thang Bui-Quoc and Dickson, J.I. Low-Cycle Fatigue Behaviour and Cumulative Damage Effect of 316 Stainless Steel at 20, 427 and 650°C.   Proceedings of Fatigue 84, Birmingham, England, in press.
20. Sinning, H.R. and Haasen, P., Cyclic Deformation of Solution Treated Cu-4 Al% Ti Single- and Polycrystals. Zeitschrift für Metallkunde, 72 (1981) 807-812.
21. Boutin, J., Marchand, N. Baïlon, J.P. and Dickson, J.I.   An Intermediate Plateau in the Cyclic Stress-Strain Curve of α-Brass.   Submitted to Materials Science and Engineering.
22. Kronmüller, H. On the Mechanism of Work-Hardening in F.C.C. Metals. Canadian Journal of Physics 45 (1967) 631-661.
23. Handfield, L. L'étude du comportement cyclique et de la fatigue oligocyclique du titane et du zirconium. Ph.D. thesis, Ecole Polytechnique de Montréal, 1984.
24. Malik, L.M. and Dickson, J.I. Yield Point Effects in α-Ti during Interrupted Tensile Tests at 300-496 K.   Materials Science and Engineering 17 (1975) 67-75.
25. Vogt, J.B., Bernard, M. and Dickson, J.I., unpublished research, Ecole Polytechnique de Montréal, 1984.

FATIGUE CRACK GROWTH - A METALLURGIST'S POINT OF VIEW

W.J. Bratina and S. Yue

Department of Metallurgy and Materials Science
University of Toronto
Toronto, Ontario, Canada M5S 1A4

ABSTRACT

Microstructural processes leading to fatigue crack formation in metals are reviewed. The eventual occurrence of persistent slip bands in polycrystalline fcc and age hardened bcc low carbon alloys is discussed. Fatigue threshold is compared to the conventional fatigue limit with particular emphasis on cryogenic behaviour. The microfractographic characteristics of near-threshold and Paris law crack propagation regions are considered with respect to striations, strongly microstructurally influenced crack growth and cyclic cleavage crack growth mechanisms. The materials under consideration are stainless steel, Ti-6Al-4V, Co-Cr-Mo, low carbon steels and HSLA steels.

1. INTRODUCTION

The early work on mechanical and microstructural changes which occur in a material subjected to cyclic straining has been reviewed by Thompson and Wadsworth (1). They concluded that the major part of a fatigue life was occupied by the growth of a crack, and they introduced the concept of a slowly growing crack with growth rates of one atomic spacing or even less. It was thus recognized that fatigue was a very localized event and that the specimen life depended on the conditions at the tip of the propagating crack. Numerous excellent reviews dealing with various aspects of the mechanisms and theories of fatigue exist (2,3,4,5). In order to avoid complications arising from the grain boundaries and from mechanical anisotropy in polycrystalline material, single crystals are frequently

used in basic fatigue studies. Single crystals of metals with an
fcc structure are particularly suitable for a number of reasons.
For example, there are no difficulties in growing large single crystals of aluminum or copper whereas there are problems associated
with growing single crystals of bcc iron or iron base alloys (6);
on the other hand niobium single crystals are relatively easily
grown by zone melting. Another reason for investigating fcc copper
is that the mechanical properties are, at best, a weak function of
temperature. Thus Basinski et al.(7) have demonstrated that the
very first stages in fatigue of copper single crystals, i.e. the
formation of persistent slip bands (PSB's) at 77 and even at 4.2 K
occur in a manner very similar to that at room temperature. Both
bcc iron and niobium, however, have mechanical properties that are
strongly dependent on temperature, the mode of deformation altering
from essentially ductile to predominantly brittle with decreasing
temperature. Finally, the plastic deformation of fcc metals is
more completely understood than the mechanisms for bcc deformation
(8). Thus the understanding of the fundamental aspects of fatigue,
particularly the early stages, have been based largely on investigations using copper single crystals, eg. Grosskreutz (9) and
Basinski (10).

In the area of metal physics, of particular interest have been
the surface changes and the changes in dislocation configurations
which occur on fatiguing a metal. In contrast, the fracture mechanics methodology has been applied primarily to studies of crack
propagation. A large body of work exists concerning the region
where the crack growth rate, da/dN is related to the stress intensity range, ΔK in an extremely simple manner and appears to obey the
Paris equation da/dN = $C(\Delta K)^n$, where C_8 and n are material constants
and da/dN values are usually above 10^{-8} m/cycle. A modified equation
to account for different R values where R = $\sigma_{min}/\sigma_{max}$ = K_{min}/K_{max}
has been proposed by Forman et al. (11). It has been stressed that
crack propagation in this region is a very weak function of microstructure at best and therefore metallurgy was not expected to contribute significantly to the understanding of Paris law fatigue
crack propagation. When, however, fracture mechanics extended its
methodology towards crack growth rates comparable to a few atomic
spacings/cycle or less, i.e. at threshold ΔK_{th}, or near-threshold
stress intensity ranges, as was suggested by Thompson and Wadsworth
(1), a large influence of microstructure was found. An improvement
of materials performance in the threshold region would be obtained,
therefore, by applying metallurgical principles to the design of
metals and alloys. The fracture mechanics approach has largely been
applied to polycrystalline commercial alloys. Work on single crystals is much less frequent (12),but there is commercial interest in
the fatigue crack propagation behaviour in single crystals of superalloys (13).

The phenomenon of fatigue is conveniently divided into the
initial changes associated with cyclic deformation (cyclic harden-

ing, softening, saturation, PSB's), followed by fatigue thresholds, crack propagation stages and final fracture (14). However, there is, of course, significant overlapping of various regions.

2. MATERIALS

The materials considered in this paper were 316L stainless steel, a Co-Cr-Mo alloy of approximately ASTM F75 composition, a Ti-6%Al-4%V alloy and a variety of bcc iron-carbon alloys including ferritic high strength low alloy (HSLA) steels. This selection of alloys was chosen to give a good representation of fcc, hcp and bcc lattices.

The structure of 316L stainless steel is austenitic, i.e. fcc. The ductility of austenitic stainless steels is such that valid K_{Ic} data cannot be obtained on specimens of reasonable size, even at cryogenic temperatures (15). Region II (the Paris law region) was always characterized by well developed fatigue striations in this material.

The structure of most commercial alloys is complex and ASTM F75 Co-Cr-Mo alloy is no different in this respect (16). Significant variations in microstructure can be achieved by various heat treatments but it remains basically an fcc structure. The tensile fracture surface at 298K is complex and there is no obvious specimen necking although there is significant elongation of the order of 5 to 15%. In our work fatigue striations were never observed in this material. This alloy is used for various body implants, particularly in orthopaedic surgery, and its success lies primarily in superior corrosion resistance properties.

Ti-6Al-4V is also extensively used as an implant material (in the extra low interstitial, ELI, condition) because of good mechanical and corrosion resistance properties. The primary importance of this alloy is, however, in aerospace because of favourable strength to weight ratio ($\sim 25 \times 10^3$ m). Below 1000°C, α(hcp) and β(bcc) phases co-exist and, although β is normally only stable at elevated temperatures, the alloy is two phase even at cryogenic temperatures. However at room temperature the alloy is predominantly α with only a few percent β. Control of microstructure depends on processing history and heat treatment and the overall effects of both can be very complex (17). As in 316L stainless steel, very distinct fatigue striations were observed in region II.

Only relatively simple iron-carbon alloys were considered. The microstructure of these alloys comprises of an α-ferrite (bcc) matrix containing well defined second phase precipitates. There is a large variation of the type of precipitates even if only the criterion of coherency and non-coherency is applied. The study of bcc iron base alloys is particularly attractive because of the ductile to brittle transition.

3. CYCLIC DEFORMATION LEADING TO FATIGUE CRACK FORMATION

Various aspects of the initial stages of fatigue including PSB's have been reviewed recently (2,5) and a large number of references dealing with cyclic stressing of copper single crystals can be found in these reviews. It is generally accepted that the occurrence of PSB's is a crucial event in the formation of fatigue cracks and this is still a subject worthy of much discussion. In single crystals PSB's can traverse the whole cross-section whereas in polycrystals the PSB's are largely confined to the surface or near surface regions. Mughrabi et al, (18) have reported in their study on fatigued fcc and bcc metals an absence of PSB's in pure α-iron and Nb single crystals.

Slip traces resembling PSB's have been reported in precipitation hardened alloys, these precipitates having a high degree of coherency with the matrix. The presence of these bands has been demonstrated in both fcc and bcc structures using TEM (19, 20, 21) and it appears that these materials are very susceptible to localized deformation in these bands. The effect of low strain ($\varepsilon = \pm 0.0014$ to ± 0.0020) high cycle (cycles to failure, $N \sim 10^6$ cycles) fatigue on the bulk and surface microstructure of polycrystalline quench aged iron-carbon alloys has been studied (19,20). Quenching and subsequent aging produced metastable, coherent carbide precipitates in the α-iron matrix. The dislocations generated during the fatigue process were observed to shear the fine carbide particles during their motion to and fro in the lattice. When the particles were reduced to a critical size by this cutting action, the particles redissolved into the matrix resulting in precipitate free channels which were easily observed using TEM. The channels observed in the bulk were parallel to the traces of {110} and {112} glide planes. Optical microscopy of the surface revealed a unique pattern of microcracks (Figure 1). The cracks were oriented mainly along two directions about 90° to one another and at 45° to the specimen axis, i.e. they were apparently related to the planes of maximum shear stress. The cracks differ from the wavy slip lines normally observed in fatigued high purity iron or iron-low carbon alloys. Subjecting this material to high strain, low cycle ($N \sim 5000$ cycles) fatigue also resulted in distinct precipitate free channels in the bulk. In this case the channels corresponded to the surface slip bands both in spacing and direction, i.e. they were parallel to traces of {110} and {112} glide planes. Near the completion of cyclic softening the channel structure transformed to a uniform dislocation cell configuration very similar to that reported by Mughrabi et al. (18) in cyclically deformed α-iron single crystals. Katagiri et al. (22) have noted crack growth occurring across the well developed dislocation cell structure in Armco iron quality polycrystalline iron. Here the cell structure was the same in the bulk and at the crack tip, and was similar to the cell structures normally observed in annealed high purity iron or in iron-low carbon alloys subjected to cyclic stressing. In a recent study Vogel et al. (21) investigated the

development of dislocation structures in unnotched and notched specimens of Al-Zn-Sn single crystals (fcc structure) which is a precipitation hardened alloy containing a fine dispersion of η' precipitates.
It was shown that the use of notched specimens permitted the observation of the development of the dislocation microstructure from the
undeformed state until the formation of PSB's. The dislocation structure in front of the tip of a fatigue crack was found to be a PSB de-
nuded of precipitates, the band length being dictated by the crack
tip stress intensity. As for the metastable carbides in the Fe-C
system discussed earlier, the narrow η' particle free PSB's were also
a consequence of particles being sheared by the to and fro dislocation motion.

In a very recent study by Gonzalez and Laird (23) on cyclic response of HSLA steel special attention has been paid to the formation of PSB's both at the surface and in the bulk of the metal.
Precipitates in this particular alloy were stable, predominantly
non-coherent niobium carbonitrides. Thus a shearing action by dislocations could not be expected, as was the case for the metastable
carbides (19,20). PSB's were observed on the surface but not in the
bulk, where a rather well defined structure of dislocation cells has
been found. An analogy can be drawn with an earlier study on Fe-1.5
%Cu alloy, where interactions of dislocations with non-coherent fine-
ly dispersed copper precipitates of 8 to 25 nm have been studied in
high strain low cycle fatigue (24). Strong interaction of dislocations with precipitates was observed in both studies, however the
cell structure in iron-copper alloy was not pronounced, perhaps
because of a rather limited amount of cyclic deformation in this particular study.

In unalloyed polycrystalline Ti, both slip bands and twins have
been associated with fatigue failure (25); in particular, twin boundaries have been observed to be favourable crack initiation sites,
the twins being present in the test piece prior to cyclic stressing
or produced during fatigue testing (26). Thus restricting twin formation results in improved high cycle fatigue strength in pure Ti
(27,28). The addition of Al to Ti appears to suppress twin formation possibly by promoting (c + a) slip as an alternative deformation mechanism (29). Therefore in Ti-6Al-4V twin formation is sup-
pressed and has never been observed to influence fatigue crack
initiation. Microcracks have been observed along slip bands in the
alpha phase (30) and at alpha-beta interfaces (31). Slow cooling of
Ti-6Al-4V from above the beta transus to room temperature results in
a structure comprising of coarse alpha plates aligned in colonies
with the beta phase in the form of an interplate layer. Crack initiation has been associated with intercolony boundaries and it has
been suggested that slip band-colony boundary interactions can
result in crack initiation (32). Such an effect can be seen in
Figure 2 where crack initiation occurred on the surface of a polished specimen; the surface microstructure was revealed by etching
after fatiguing.

4. FATIGUE THRESHOLD AND NEAR-THRESHOLD

4.1 Threshold Versus Fatigue Limit

A comparison of threshold stress intensity, ΔK_{th}, (4,33-38,40, 41,45) with the well established fatigue or endurance limit seems inevitable because, although there appear to be many differences between the concepts behind these two values, the primary connection, i.e. initiation, appears to be very strong. Fatigue limit, which has been observed in bcc metals, particularly in bcc iron-carbon alloys and in many ferritic iron-base alloys, seems to be related to the interactions of dislocations and interstitials. The process is essentially diffusion controlled and Snoek-Schoeck-Seeger and Cottrell-Bilby mechanisms operate (42). A number of fatigue characteristics of bcc metals and alloys appear to be satisfactorily explained by these dislocation pinning mechanisms or strain aging (1). Dickson et al. (43) recently used strain aging in order to explain an increase of the threshold values for a water quenched low C steel as compared to an air cooled material. However, the interpretation of threshold behaviour on the basis of microstructural parameters is not always satisfactory and the effect of environment must be considered. Environment is known to have an influence on the fatigue behaviour of smooth, mirror polished specimens (44), but the effect of environment on threshold is far more significant. Inert atmospheres or vacuum are now used frequently in threshold determinations and at 77K immersion in liquid nitrogen is a standard procedure.

The effect of cryogenic temperatures on ΔK_{th} of iron-base alloys exhibiting ductile to brittle transition in the range of temperatures of interest has been investigated by Gerberich and co-workers (33,36, 38,40,45). In some steels such as many niobium containing low-carbon HSLA steels, however, a completely ductile dimpled structure was observed even at 77K when fractured in a tensile manner, whereas steels of similar compositions and microstructures but with a higher volume fraction of iron carbide precipitates due to slightly higher carbon contents exhibited predominantly brittle fracture at 77K and a mixed mode of fracture between 77 and 298K (46). Thus there exist two widely differing cases, based on the effect of temperature on fracture, which must be considered. In Figure 3, the effect of temperature on expected crack propagation behaviour of the group of steels which retain a ductile fracture appearance down to 77K is shown schematically. A shift of K_{Ic} to lower values should occur in accordance with the K_{Ic} versus temperature relationship. Thus a very large increase in the exponent, n of the Paris equation is expected (36; compare also Fig. 4.35 for steel A533B in Ref. 45) assuming threshold is independent of temperature. However, a large, in some cases up to 40 percent, increase of ΔK_{th} has been observed at 77K compared to room temperature data (37). This effect was initially attributed to environmental factors; but it now appears to be structure dependent. There have been attempts to relate changes in threshold with decrea-

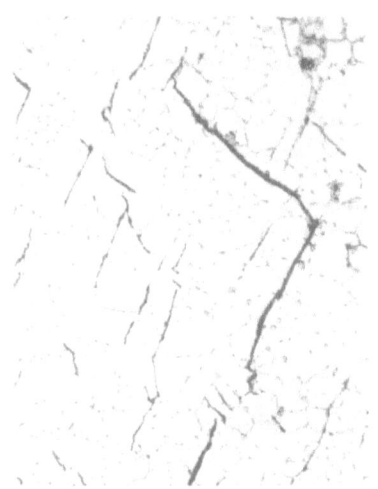

Fig 1:- Fatigue microcracks on the surface of a quenched and aged Fe-C alloy(electro-polished and etched after fatiguing).

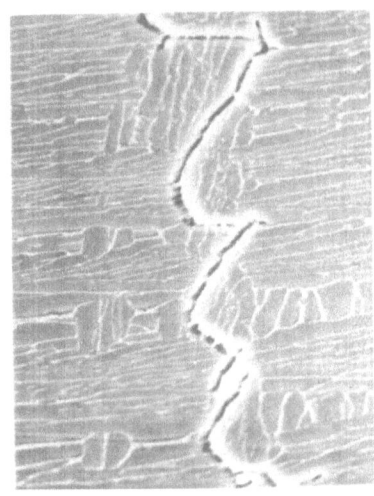

Fig 2:- Surface crack at colony boundaries of fatigued lamellar Ti-6Al-4V alloy.

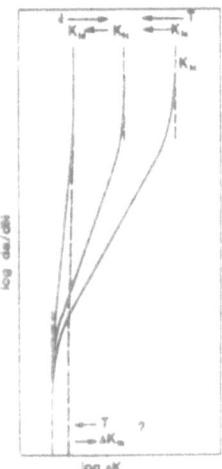

Fig 3:- Expected variation of crack propagation curves approaching cryogenic temperatures.

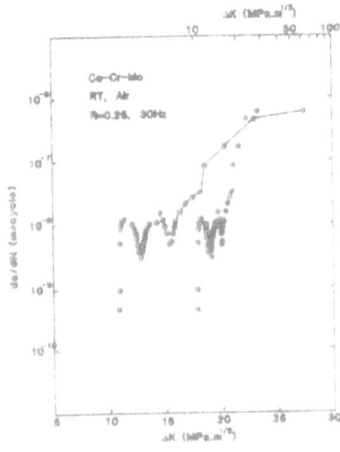

Fig 4:- Fatigue crack propagation curve for Co-Cr-Mo alloy(47).

sing temperature to the increase in tensile strength which occurs in bcc metals with decreasing temperature. However, in that case the threshold temperature dependence would parallel that of fatigue limit. On the basis of defect structure of the specimen surface, the opposite behaviour would be expected at cryogenic temperatures. An eventual ductile to brittle transition as experienced by many ferritic steels, including some HSLA steels, appears to complicate the da/dN versus ΔK trend at cryogenic temperatures even further (33, 36, 38). Apparently more research on microstructurally well characterized bcc metals and alloys is needed before all the intricacies of the near-threshold crack growth at cryogenic temperature can be explained in terms of basic deformation mechanisms.

4.2 da/dN Measurements

The metallurgical understanding of crack propagation at ultra low rates is such that it appears that threshold crack growth does not occur gradually or continuously, as a da/dN curve might indicate, but in irregular steps as suggested by studies of fracture surfaces using SEM. However, practically any experimental method is expected to be unsatisfactory in resolving the very fine microstructural details of near-threshold crack growth. In this respect a consideration of the fracture of Co-Cr-Mo, for example, may help to understand the near-threshold fracture characteristics. In this alloy the fracture mode remains predominantly crystallographically influenced throughout the fatigue life up to final fracture. The respective da/dN vs ΔK curve is shown in Fig. 4 (47); da/dN is plotted against linear and logarithmic ΔK values for clarity. Even with crack length detection equipment of resolution far inferior to that of SEM, Fig. 4 still illustrates the step-like progress of the propagating fatigue crack. However, the possibility of studying such behaviour at threshold levels is extremely small. But there is little doubt that irregular steps are the predominant mechanism by which cracks can progress in the ultralow crack growth regime.

4.3 Fractographic Studies

In the past we have performed a large number of fractographic studies on fatigues 316L stainless steel, Co-Cr-Mo (ASTM F75) and Ti-6Al-4V ELI alloys. Both SEM and TEM replica techniques were extensively employed resulting in a maximum resolution of the order of 5×10^{-8}m, i.e. a striation spacing of 50nm could be easily resolved. Thus growth rates of the order of 10^{-8} or at best 10^{-9}m/cycle, if a correction is made for tortuosity as will be explained later, can be observed. This is orders of magnitude higher than the crack propagation rates encountered in the threshold region; thus it is difficult to analyse this region even on the basis of high resolution fractographs. For metals which exhibit fatigue striations on the fracture surface, striations can be resolved at the higher end of the near-threshold crack propagation values. Their direction is in general perpendicular to the crack propagation direction. If, in this region the direction of striation-like traces changes from grain to grain, it may be that these are slip plane traces associated with stage I deformation. In stage I, the crack is expected to propagate along crystallographic slip planes which experience maximum shear stress. This stage can then be regarded as an extension of the crack nucleation process which occurs via PSB's in single crystals, on the surface of polycrystalline fcc materials or structures which behave in a similar manner to fcc metals. Thus cracks are expected to propagate in PSB's, once these have been formed. Figure 5 shows the near-threshold region of a Ti-6Al-4V alloy with a lamellar microstructure as illustrated in Fig. 2. It seems unlikely that the relatively large striated regions are fatigue striations. At higher ΔK values the fatigue striations in this alloy are clearly resolved and it therefore appears that the striated regions in Fig. 5 may be slip plane traces. Figure 6 is typical of the near-threshold region in 316L stainless steel, but it is not very revealing. Fatigue fracture of Co-Cr-Mo at near-threshold is shown in Fig. 7. As mentioned previously, fatigue striations were never observed in Co-Cr-Mo and crack growth was almost completely crystallographically controlled. At low magnifications, the region illustrated in Fig. 7 was revealed as a featureless brittle flake fracture. The higher magnification, however, reveals a strongly crystallographic structure indicating step-like crack propagation and implying non continuous crack growth.

5. FATIGUE CRACK PROPAGATION

Fatigue crack growth in Region II can proceed in polycrystalline materials by various deformation modes. Striations are common in ductile materials such as 316L stainless steel, Ti-6Al-4V and many low carbon steels, at room temperature at least. Microstructurally influenced propagation following crystallographically oriented planes is observed in alloys with pronounced heterogeneous microstructures such as Co-Cr-Mo. Large brittle facets with cleavage steps, cleavage rivers and some features resembling widely spaced brittle striations have been observed in some bcc iron-carbon alloys even at very high magnifications. Many experiments have been performed at ambient atmosphere and the behaviour in vacuum may differ. However, the results on strongly corrosion resistant alloys such as Co-Cr-Mo or results obtained at cryogenic temperatures are expected to be largely similar to results obtained in vacuum and therefore representative of the fracture characteristics of the structures under investigation.

5.1 Striations

Using direct observations of striations on fatigue fractured surfaces, it is perhaps possible to cover two orders of magnitude of crack growth rate. On the other hand, da/dN values obtained by macro-crack growth monitoring may be extended easily over five, or even more, orders of magnitude and to the threshold values far beyond the power of any microscope. It is usually assumed that each striation corresponds to a crack advance per cycle and therefore a striation

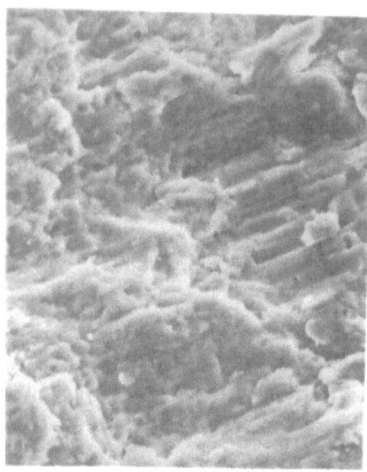

Fig 5 :- Near threshold
fracture surface of lamellar
Ti-6Al-4V.

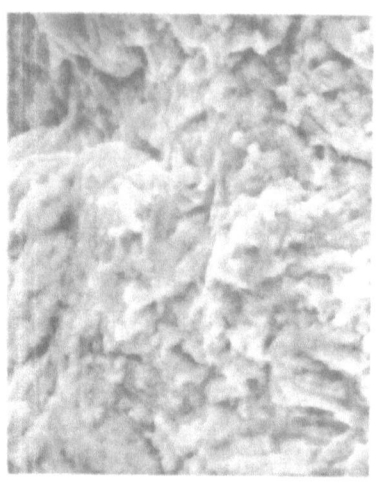

Fig 6:- Near threshold
fracture surface of 316L
stainless steel.

Fig 7:- Crystallographically
controlled fracture at near-
threshold in Co-Cr-Mo. Note
fine steps.

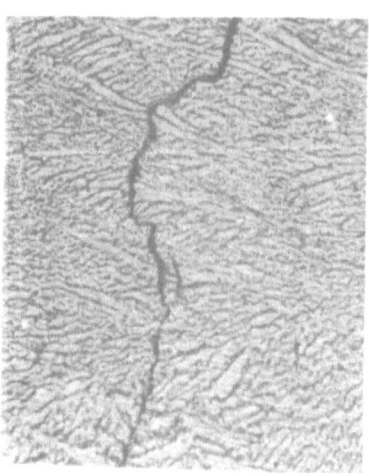

Fig 8:- Crack path tortu-
osity in lamellar Ti-6Al-4V.

spacing should be identical to da/dN calculated from macro-crack growth observations versus number of cycles for the same ΔK. Such correspondence has been occasionally demonstrated but it may be purely coincidental. In principle the values of striation spacing and da/dN should differ, the striation spacing should be larger than the corresponding da/dN. This is easily appreciated when an actual crack path is observed, Fig. 8. The path is tortuous and therefore always longer than the straight path postulated by fracture mechanics and used in da/dN calculations. Very accurate measurements of striation spacings in 316L stainless steel yielded up to one order of magnitude higher values than da/dN measured macroscopically, Fig. 9. However, a similar discrepancy at threshold level has been explained by discontinuous crack growth at these ultralow propagation rates (37).

In controlled fatigue experiments on 316L stainless steel specimens accurate striation counts were performed using SEM, and, at lower propagation rates, TEM on replicas. In all cases resolvable striations accounted for only 5 to 10% of the total fatigue life. Final fracture is a relatively fast process and the contribution of this stage to the total fatigue life can be disregarded in these experiments. Therefore 90 to 95% of the fatigue life must be spent in crack initiation and threshold processes. If the concept that the fatigue process is largely a crack growth process is assumed, then crack initiation is only a small fraction of the total life and 80% or even more of fatigue life is spent in the threshold or near-threshold regions. The importance of the threshold region therefore cannot be over-emphasized when design is based on crack growth criteria. However, if design is based on fatigue endurance or limit, crack initiation should not occur at all.

Fatigue striation spacing is highly sensitive to the loading pattern and, using striation characteristics obtained on laboratory specimens, fracture stresses and stress patterns can be determined for specimens which failed under unknown loading circumstances. In particular we investigated 316L stainless steel implants where a direct stress analysis was extremely difficult. By comparing striation characteristics of laboratory fatigued specimens to actual implant failures, Fig. 10, a large body of information has been obtained on load magnitudes and random load variations to which these implants are subjected in the human body (48,49). Bates and Clark have related striation spacing to $(\Delta K/E)^2$ and this has been occasionally utilized in such investigations but its usefulness appears to be limited. Quantitative stress determinations via striation characteristics are, of course, limited to alloys where striations are a prominent microstructural feature. However in similar load conditions the information obtained from materials exhibiting striations should apply to any material, thus making design improvements possible.

5.2 Crack Path Strongly Influenced by Crystallography

When a fatigue crack advances by deformation modes influenced strongly by microstructure, fractography and fracture path observations reveal distinctly oriented facets and a complete absence of striations (fig. 11a, b). Many commercial alloys behave in a similar manner although sometimes a few faint striation-like features have been observed (50). Even though the fracture appearance resembles that of quasi-cleavage, crack growth occurs in all probability along planes of localized slip. However, some cleavage fracture cannot be excluded. In single crystals of some superalloys a similar pattern of crack propagation has been observed, the cracks tending to propagate on what appears to be crystallographic planes for relatively long distances (51). In a complex γ-Fe-Ni-Al polycrystalline alloy, a zig-zag fatigue crack path was observed occurring by localized planar slip on two intersecting slip planes (52) resulting in fracture characteristics similar to those shown in Fig. 11 a and b. There are many indications that grain size-plastic zone interactions dictate the fracture characteristics and the shape of the da/dN vs ΔK curves of the metal under test. For example, the grain size, as defined in metallurgical terms, of the Co-Cr-Mo alloy is unusally large, of the order of millimeters, whilst that of HSLA steels is a few μm. Ti-6Al-4V in the beta annealed condition has a very coarse prior beta grain size comparable to the grain size of the Co-Cr-Mo alloy, a much smaller colony size, and the size of the plates which make up the colonies are of the order of a few μm. Whatever the controlling microstructural feature of Ti-6Al-4V or Co-Cr-Mo, two completely different fracture patterns are observed in stage II.

5.3 Crack Growth in Brittle Materials

Many commercial very high strength alloys tend to fracture in a brittle manner even at room temperature. At cryogenic temperatures many ferritic (bcc) alloys change from ductile mode of deformation and fracture to a brittle mode. There is also a monotonic strength increase associated with temperature decrease but this is not necessarily related to fracture mode (46). It is interesting to note that the Paris law exponent increases in a similar manner as the monotonic effective plastic flow stress with decreasing temperature (36). This similarity may be more than merely coincidental. Quantitative fractography of cyclic cleavage crack growth features is a difficult task but very recently an attempt has been made by Gerberich et al. to rationalize the observations of the behaviour of iron and iron binary alloys at cryogenic temperatures in terms of a dislocation model (33, 39). Fatigue crack growth investigations at cryogenic temperatures in various bcc metals are certainly of interest since both plastic flow stress and fracture mode are strongly temperature dependent.

ACKNOWLEDGEMENT

The financial support of the Natural Sciences and Engineering Research Council of Canada is gratefully acknowledged.

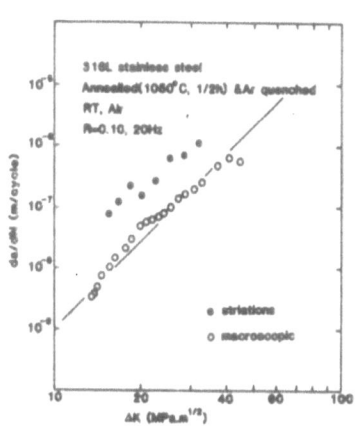

Fig 9:- Comparison of crack propagation curves determined by macro and microscopic(SEM-TEM)methods.

Fig 10:-Striations in 316L stainless steel hip implant fractured"in vivo"; region II(Paris Law region).

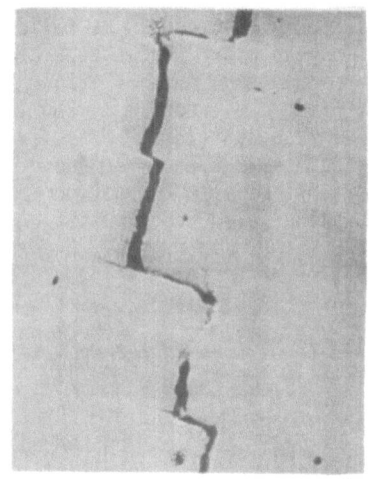

Fig 11a

Fig 11b

Crystallographic control of crack growth in the Paris Law region as shown on the fracture surface(fig11a)and the crack path(fig 11b).

REFERENCES

1. Thompson,N. and J.N. Wadsworth,Metal Fatigue,Advances in Physics 7 (1958) 72-170.
2. Laird,C.,Mechanisms and Theories of Fatigue, Fatigue and Micro-structure (ASM Seminar,1979) 149-203.
3. Starke,E.A. and G. Lutjering, Cyclic Plastic Deformation and Microstructure, ibid. 205-239.
4. Fine, M.E. and R.O. Ritchie, Fatigue Crack Initiation and Near-Threshold Crack Growth, ibid. 245-278.
5. Mughrabi, H., Microscopic Mechanisms of Metal Fatigue,Strength of Metals and Alloys,Proc.5th Int.Conf.Vol.3,Pergamon(1980)1615-1637.
6. Rosinger, H.E.,W.J. Bratina and G.B. Craig,Growth of Cylindrical Iron Single Crystals by the Strain-Anneal Technique, J. of Crystal Growth 7 (1970) 42-44.
7. Basinski,Z.S., A.S. Korbel and J.S. Basinski,Acta Metall. 28 (1980) 191.
8. Christian, J.W.,Some Surprising Features of the Plastic Deforma-tion of BCC Metals and Alloys,Metall.Trans.14 A (1983) 1237-1256.
9. Grosskreutz,J.C.,Fundamental Knowledge of Fatigue Fracture, Int. Congress on Fracture (Munich, 1973) PLB-212.
10. Basinski, Z.S.,R. Pascual and S.J. Basinski, Low Amplitude Fatigue of Copper Single Crystals,Acta Metall. 31 (1983) 591-602.
11. Forman,R.G., V.E. Kearney and R.M. Engle,Trans. ASME 89 (1967)459.
12. Neumann,P., H. Fuhlrott and H. Vehoff, Experiments Concerning Brittle, Ductile and Environmentally Controlled Fatigue Crack Growth, Fatigue Mechanisms, ASTM STP 675 (1979) 371-395.
13. Runkle, J.C. and R.M. Pelloux, Micromechanisms of Low-Cycle Fati-gue in Nickel-Based Superalloys at Elevated Temperatures, Fatigue Mechanisms, ASTM STP 675 (1979) 501-527.
14. Fatigue 84 Programme. 2nd Int. Conf. on Fatigue and Fatigue Thresholds (1984), University of Birmingham.
15. Campbell, J.E., Fracture Properties of Wrought Stainless Steels, Applications of Fracture Mechanics, ASM (1982) 105-167.
16. Kilner, T.,Ph.D. Thesis, University of Toronto (1984).
17. Rosenberg, H.W., J.C. Chesnutt and H. Margolin, Fracture Proper-ties of Titanium Alloys, Applications of Fracture Mechanics, ASM (1982) 213-252.
18. Mughrabi, H., F. Ackermann and K. Herz, Persistent Slip Bands in Fatigued FCC and BCC Metals,Fatigue Mechanisms,ASTM STP 675 (1979) 69-105.
19. McGrath, J.T. and W.J. Bratina, Interaction of Dislocations and Precipitates in Quench Aged Fe-C Alloys Subjected to Cyclic Stressing, Acta Metall. 15 (1967) 329-339.
20. McGrath, J.T. and W.J. Bratina, The Mechanical and Microstructural Changes in Quench-Aged Fe-C AlloysSubjected to Cyclic Straining, Czech. J. Physics B19 (1969) 284-293.
21. Vogel, W., H. Wilhelm and V. Gerold, Persistent Slip Bands in Fatigued Peak Aged Al-Zn-Sn Single Crystals, Acta Metall. 30 (1982) 21-30.

22. Katagiri,K., J. Awatani, A. Omura, K. Koyanagi and T. Shiraishi, Dislocation Structures Around the Crack Tips in the Early Stage in Fatigue of Iron, Fatigue Mechanisms, ASTM STP 675 (1979)106-128.

23. Gonzalez, G. and C. Laird, The Cyclic Response of Dilute Iron Alloys, Metall. Trans. 14 A (1983) 2507-2515.

24. McGrath, J.T. and W.J. Bratina, Fatigue of an Fe-1.5% Cu Alloy Containing Stable, Non-coherent Precipitate Particles, Phil. Mag. 21 (1970) 1087-1091.

25. Conrad, H., M. Doner and B. de Meester, Deformation and Fracture, Titanium Science and Technology, Vol. 2, (Jaffee and Burte, eds. Plenum Press, 1973) 969-1005.

26. Partridge, P.G. and C.J. Peel, Effect of Cyclic Stress on Unalloyed Polycrystalline Titanium, The Science, Technology and Application of Titanium (Jaffee and Promisel eds. Pergamon Press, 1970) 517.

27. Beevers, C.J., Fatigue Behaviour of Alpha Titanium and Alpha-titanium-hydrogen Alloys, ibid. 535.

28. Beevers, C.J. and J.L. Robinson, Some Observations on the Influence of Oxygen Content on the Fatigue Behaviour of Alpha Titanium, J. Less Common Metals 17 (1969) 345-352.

29. Benson, D.K., J.C. Grosskreutz and G.G. Shaw, Mechanisms of Fatigue in Mill Annealed Ti-6Al-4V at RT and 600°C,Metall. Trans. 3A (1972) 1239.

30. Wells, C.H. and C.P. Sullivan, Low Cycle Fatigue Crack Initiation in Ti-6Al-4V, Trans. ASM 62 (1969) 263.

31. Stubbington, C.A. and A.W. Bowen, Improvements in the Fatigue Strength of Ti-6Al-4V Through Microstructural Control, J.Materials Science 9 (1974) 941-947.

32. Brown, R. and G.C. Smith, Crack Initiation in an Aligned Titanium Alloy Containing an Interface Layer, Scripta Metall.15 (1981) 357-360.

33. Gerberich, W.W., W. Yu and K. Esaklul, Fatigue Threshold Studies in Fe , Fe-Si and HSLA Steels, Parts I and II, Metall. Trans. 15A (1984) 875-900.

34. Ritchie, R.O., Near-Threshold Fatigue Crack Propagation in Steels, International Metals Rev. No. 245 (1979) 205-230.

35. Dickson, J.I., J.P. Bailon and J. Masounave, A Review on the Threshold Stress Intensity Range for Fatigue Crack Propagation, Can.Met. Quarterly 20 (1981) 317-329.

36. Gerberich, W.W. and N.R. Moody, A Review of Fatigue Fracture Topology Effects on Threshold and Growth Mechanisms, Fatigue Mechanisms, ASTM STP 675 (1979) 292-341.

37. Tschegg, E. and S. Stanzl, Fatigue Crack Propagation and Threshold in bcc and fcc Metals at 77 and 293K, Acta.Metall.29 (1981) 33-40.

38. Gerberich, W.W. and K.A. Peterson, Micro and Macromechanics Aspects of Time Dependent Crack Growth, Micro and Macro Mechanics of Crack Growth, AIME The Metall.Society (1982) 1-17.

39. Gerberich,W.W. and K. Jatavallabhula, Quantitative Fractography and Dislocation Interpretations of the Cyclic Cleavage Crack Growth Process, Acta Metall. 31 (1983) 241-255.

40. Yu, W. and W.W. Gerberich, On the Controlling Parameters for Fatigue Crack Threshold at Low Homologous Temperatures, Scripta Metall. 17 (1983) 105-110.
41. Fatigue Thresholds, International Conference, Stockholm, Backlund, J., A.F. Bloom and C.J. Beevers, eds. (1982) Various papers.
42. Rosinger, H.E., G.B. Craig and W.J. Bratina, The Recovery of Internal Friction in an Iron-Carbon Alloy, Phil. Mag. 25 (1972) 1331-1343.
43. Dickson, J.I., M-C. Lu and J.P. Bailon, The Influence of Strain Aging on the Fatigue Crack Propagation Threshold, Scripta Metall. 17 (1983) 49-52.
44. Majundar, D. and Y-W Chung, Surface Deformation and Crack Initiation During Fatigue of Vacuum Melted Iron:Environmental Effects, Metall. Trans. 14 A (1983) 1421-1425.
45. Gerberich, W.W., R.H. Van Stone and A.W. Gunderson, Fracture Properties of Carbon and Alloy Steels, Appl. Fracture Mechanics, ASM (1982) 41-103.
46. Yue, S., and W.J. Bratina, The Deformation Behaviour of a HSLA Steel Over the Temperature Range from 77 to 298K, to be published.
47. Kannan, A., M.A.Sc. Thesis, University of British Columbia (1982).
48. Bratina, W.J. and R.M. Pilliar, Fatigue Characteristics of Metallic and Non-Metallic Surgical Implant Materials, Proc. 1st Mediterranean Conf. Biomedical Eng., Sorrento, Italy, (1977) Vol. II, 28.
49. Bratina, W.J., A.C. Wallace and J.L. Co, Fracture Studies of Stainless Steel Orthopaedic Implants, Proc. 3rd Mediterranean Conf. Biomedical Eng., Portorož, Yugoslavia, (1983) 2.1.
50. SEM and TEM Fractography Handbook, Battelle Memorial Institute, Columbus (1975) and Electron Fractography Handbook, Battelle Memorial Institute, Columbus (1976).
51. R.H. Jeal, Rolls Royce, U.K. Private communication.
52. Hornbogen, E. and K. H.Zum Gahr, Acta Metall. 24 (1976) 581-592.

THE EFFECT OF SOLUTE DISTRIBUTION ON CREEP FRACTURE IN A Ni ALLOY

G. Burger and D.S. Wilkinson

Institute for Materials Research
McMaster University
Hamilton, Ontario, Canada
L8S 4M1

ABSTRACT

Creep cavitation is a heterogeneous process. Thus, cavities coalesce locally to form microcracks long before final fracture. This process has been studied in a Ni-1% Sn alloy whereby macrosegregation of the tin during casting produces an inhomogeneous distribution of cavities and microcracks. The result is an increase in both strength and ductility, as compared to the same alloy with a homogeneous distribution of solute. This is because, while cavities and microcracks form readily in the Sn-rich regions, failure requires the propagation of cracks through regions with low Sn content where the creep ductility is high.

INTRODUCTION

A great deal of effort, both experimental and theoretical, has been expended in attempting to understand the process of grain boundary creep fracture. It is clear that the process involves the nucleation, growth and coalescence of grain boundary cavities. It is generally believed (1,2,3) that cavities initiate at inhomogeneities in the grain boundary, such as jogs or precipitates, or at grain boundary triple-points. Grain boundary sliding, grain boundary diffusion, surface diffusion and power-law creep are all thought to play a role in cavity nucleation and the subsequent growth of cavities. Models for each of these mechanisms have been developed which predict the kinetics and their dependence on stress and material and microstructural parameters (2-6).

The application of these models to predict the creep life of materials requires the assumption that the material is homogeneous on a scale larger than the grain size, or the average precipitate spacing. However, many of the materials used in high temperature service are inhomogeneous in both chemistry and microstructure over dimensions much larger than this, the result of the specific processing history. For example, there is a class of superalloys, used in the as-cast condition, in which cavitation may be highly inhomogeneous (7). The evolution of damage in such heterogeneous materials has been modelled by Dyson (8,9). He assumes that cavitation, occurring preferentially in a local region, is constrained by the surrounding material to creep at the macroscopic creep rate. This constraint is the result of the dilation produced by cavity growth. In using this treatment to predict fracture, the size of the cavitating region is first defined. The time (or strain) required for cavities to coalesce in this region is then calculated (10). Again, actual material behaviour differs from this model (11), in that the "zones" of damage tend to grow with time. The possible interaction between damaged regions also needs to be taken into account.

Thus, the quantitative analysis of cavitation has not been adequately treated, due in part to a lack of quantitative data on creep damage development. In the present study, we have investigated the accumulation of damage in a Ni-1% Sn alloy. The influence of tin additions in promoting creep brittleness in nickel has been reported by White and Padgett (12). However, the segregation of tin during casting can result in cavitation developing preferentially in local regions. The results obtained in these tests are compared with those on an alloy of similar composition, but with a more uniform solute distribution (13).

EXPERIMENTAL

Pure (99.99%) nickel was alloyed with (nominally) 1% tin. The alloy was sand cast under vacuum. It was then swaged through a total reduction in area of 86%, with intermediate anneals at 800°C for one hour between passes. Samples were then machined to a gauge length of 25.4 mm and diameter of 5 mm. They were annealed at 1000°C for one hour in high vacuum prior to testing.

The cast dendritic structure was not completely removed during thermo-mechanical processing. As a result, the specimens contained elongated regions of material, aligned with the tensile axis of the samples, in which the grain size was much smaller than the average grain size of 86 μm. Electron microprobe analysis revealed these clusters to be interdendritic regions of high solute concentration. Although the material was heterogeneous on the scale of the interdendritic spacing, chemical analysis on several samples showed the average composition was constant with a tin concentration of 1.30 ± 0.01 wt%.

Creep tests were performed in a Dennison constant-load creep machine, under an atmosphere of Ar/10 H_2 at 5 p.s.i.g. All tests were at 600°C, with a maximum variation of 2°C over the sample gauge length. Three stress levels were investigated, nominally 113, 142, and 179 MPa. Several tests were performed at each stress, some of which were interrupted prior to failure, so that the damage accumulation process could be examined in detail.

Specimens were sectioned along the tensile axis for metallographic examination. An etch-polishing technique was used to remove material which smeared over the cavities during grinding. Extensive regions of the sample were photographed, at a magnification of 160X (as this was the minimum magnification which provided all the significant information on the accumulation of damage). Histograms of cavity and crack size distributions were obtained using a digitizing tablet interfaced with a microcomputer. The cavity or crack size was defined as the maximum projected length normal to the tensile axis.

RESULTS

The creep curves for the three stress levels are plotted in Fig. 1, while the data obtained from these is summarized in Table I. Included in this table are results obtained from the work of White et al. (13), in which a similar investigation on a chill-cast Ni-1% Sn alloy was conducted.

Table 1

	σ(MPa)	Failure time t_f(sec)	$\log t_f$	Elongation to failure ε_f	ε_{mcr}
	113.3	6.75×10^5	5.83	0.234	2.12×10^{-7}
Present	113.3	7.66×10^5	5.88	$\simeq 0.24$	2.03×10^{-7}
work	141.6	1.77×10^5	5.25	0.32	1.09×10^{-6}
	179.4	3.26×10^4	4.51	0.360	6.30×10^{-6}
	179.4	3.14×10^4	4.50	0.338	6.06×10^{-6}
Data from	34.5	7.4×10^6	6.9	0.065	5.8×10^{-9}
White et al.	60	5.7×10^5	5.8	0.071	9.9×10^{-8}
	103.5	7.0×10^4	4.8	0.078	5.9×10^{-7}

Figure 2 shows how the strain rate varies with strain for one stress level. No true steady-state is attained because of the constant load conditions. However, the minimum creep rate ε_{mcr} can be determined. The strain at which this occurs increases with stress (Fig. 3). Figure 4 reveals a power-law creep relationship for our results, and for those of White et al. (13). However, both the creep strength and stress exponent n are different. Our material exhibits an n value of 7.2, while White et al. obtain a value of 4.2. At a comparable strain rate, our material is stronger. The elongation to failure in our material is also significantly greater, even when compared at the same stress. Thus, at stress levels near 100 MPa, the material used in the present study is both stronger and more ductile than that used by White et al.

The level of damage was measured on polished sections taken from the interrupted test samples. Grain offsets at the surface due to boundary sliding are observed by the end of primary creep. However, no internal cavitation is evident at this point, using magnifications up to 640X. It is only after the minimum creep rate is attained that internal damage is detected. Even then, damage accumulation is slow. Both wedge-cracking and r-type cavities are observed. The propagation of wedge cracks occurs by linking with cavities growing ahead of the crack, as shown in Fig.5.

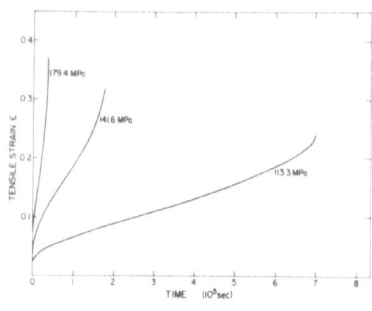

Fig. 1 — Tensile creep strain vs. time at 600°C.

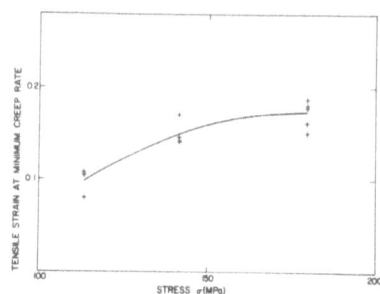

Fig. 3 — Variation of strain at minimum creep rate with stress.

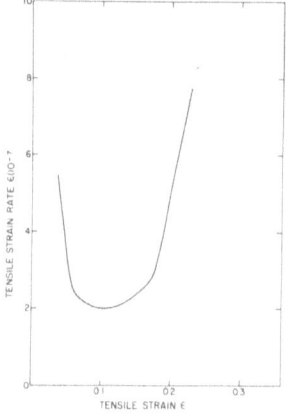

Fig. 2 — Tensile strain rate vs. strain at 113.3 MPa.

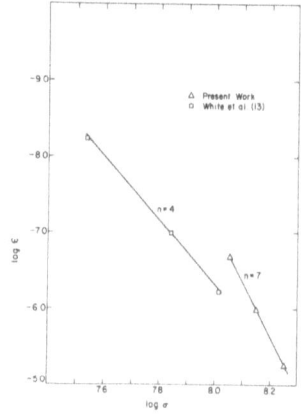

Fig. 4 — Plot of log ε_{mcr} vs. log σ.

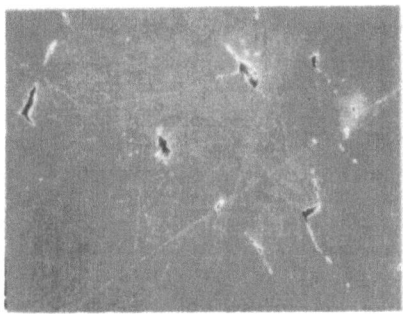

Fig. 5 – Photograph showing wedge cracks propagating
by linking with voids ahead. (90X)

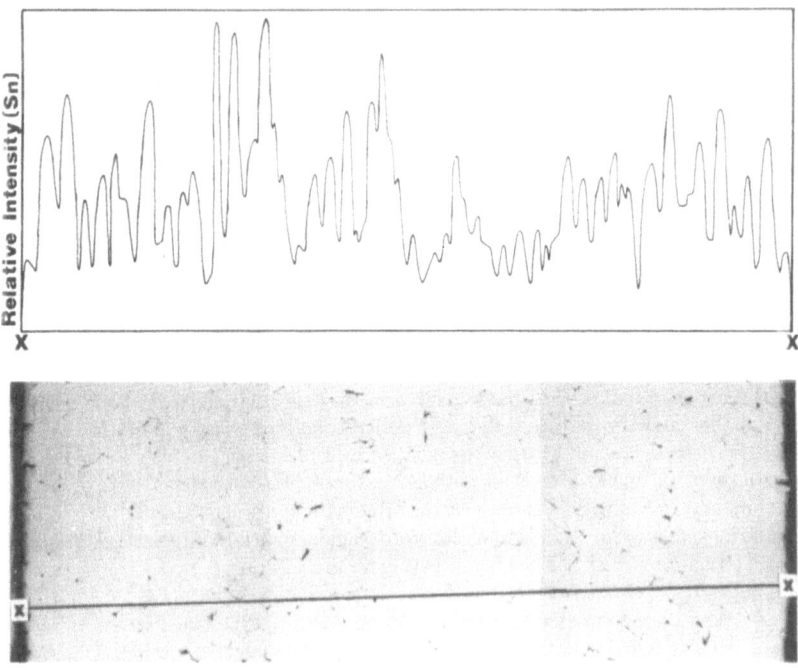

Fig. 6 –   Photogaph (26X) showing damage localized in regions aligned with the
tensile axis (vertical). The microprobe scan along X-X for Sn (the
intensities are relative to background) indicates that these regions are
associated with higher levels of Sn.

Damage is initially concentrated in regions where the grain size is several times smaller than average. Microprobe scans across the sample (see Fig. 6) revealed that these regions contain high concentrations of tin, several times the average level in the sample. Extensive cavitation and microcracking within these regions occurs before any appreciable damage is found elsewhere. In some instances, microcracks as large as the damage zone itself are observed. The level of damage accumulated in the regions separating the damage zones is relatively low when final fracture occurs.

Examination of the fracture surfaces reveals that, although the fracture appears to be intergranular, only small regions are cavitated. The remainder of the fracture surface is characterized by a rippled appearance with much less cavitation. The final fracture of the region between the damage zones appears to involve little cavitation. In addition to the cavitation which develops internally, significant amounts of surface cracking are also present.

Damage Analysis

The level of damage was measured over a region running across the entire sample cross-section and extending along the sample length for a distance of approximately the sample diameter. The variation of the damage distribution with interdendritic spacing could then be taken into consideration. The extent of surface cracking was also measured. Under some conditions, these interacted with internal cracks and cavities and appeared to be significant in determining the final fracture. In an initial attempt to characterize the damage levels, cavity size distributions were obtained for various stages in the creep process. Data for internal and surface damage were analyzed separately. Only the conclusions drawn from this analysis will be given in this paper.

Figure 7 shows the density of cavities as a function of strain. It is clear that cavities nucleate throughout most of the creep life. Indeed, the actual nucleation rates are higher than indicated by the measured increase in cavity density, since many of the smaller cavities coalesce with larger cavities and microcracks. Near failure, the cavity density decreases, also due to coalescence. The cavity density at a given strain also depends on stress, increasing with decreasing stress. In contrast with this data, we find that the density of surface cracks remains roughly constant with strain over the strain levels examined. Most of the grain boundary/surface intersections result in crack formation, which appears to nucleate early in creep life.

The average internal damage level is shown as a function of strain in Fig. 8. Here, damage is defined as the fraction of the sample diameter cavitated. This value is obtained by summing the projected cavity lengths normal to the tensile axis ℓ_c, and dividing by the total projected grain boundary length normal to the axis in the area sampled, ℓ_g. The length ℓ_g is set equal to $n_g D$, where D is the average grain size and n_g is the number of average size grains in the area sampled. As with the cavity density, the damaged area fraction is both strain- and stress-dependent. Very little cavitation is found at strains just greater than that at the minimum creep rate, so that it appeared that the onset of significant damage accumulation coincides with the development of a slight neck.

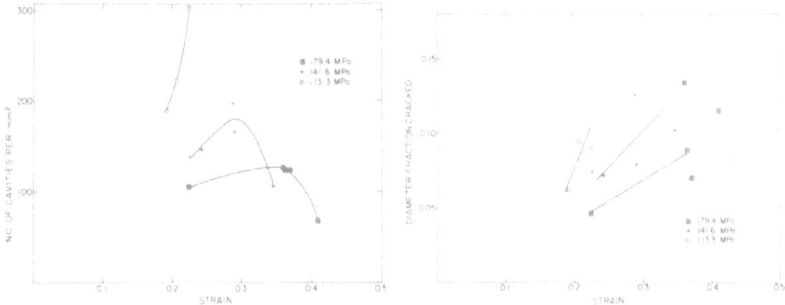

Fig. 7 — Plot of cavity number density vs. strain.

Fig. 8 — Plot of average internal damage level vs. strain, expressed as the fraction of diameter which is cavitated.

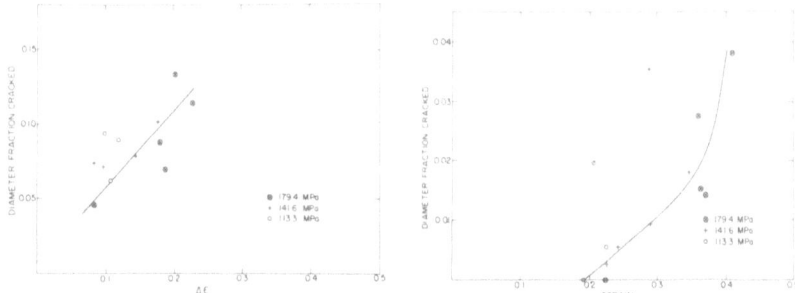

Fig. 9 — Plot of average internal damage level vs. strain Δε beyond the minimum creep rate.

Fig. 10 — Average internal damage level vs. strain, where the damage is associated with cracks larger than the average grain size

When the damage is plotted against the tertiary strain (i.e., the strain beyond that at the minimum strain rate) Δε, as in Fig. 9, the stress dependence is reduced. However, the data still show considerable scatter.

We also measured the average fraction of the sample diameter covered with surface cracks. The same dependence on stress and strain is evident. However, the average amount of surface cracking increases more rapidly with strain at lower stresses, due to the development of a large dominant surface crack, a feature not observed at higher stresses.

The average fraction of the specimen diameter which is covered with internal cracks larger than the average grain size is shown in Fig. 10. This type of damage is a

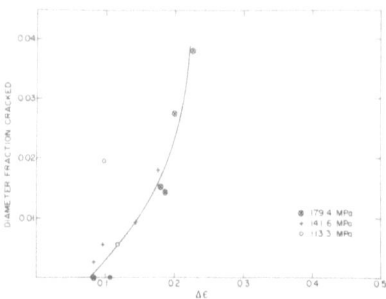

Fig. 11 – Average internal damage level vs. strain
beyond minimum creep rate, where
damage is associated with cracks larger
than the average grain size.

strong function of strain. There is also, however, a weak dependence on stress. When plotted against $\Delta\varepsilon$, the tertiary strain (Fig. 11), this disappears. There is a minimum strain below which no cracks larger than the average grain size are found. This strain is also independent of stress.

DISCUSSION

In this study, we have analyzed in some detail the accumulation of damage in an inhomogeneous alloy. The creep curves (Fig. 1) we obtain are qualitatively similar to those obtained for a homogeneous alloy. One might expect however, that primary creep will occupy a greater proportion of life for an inhomogeneous material, due to the need for long range stress redistribution in the material. In the present case, bands of high Sn content and small grain size extend parallel to the tensile axis. Therefore, in order to maintain strain rate compatibility, stresses rise in the high strength (high Sn) regions and fall in the low strength regions.

The onset of observable cavitation coincides with the end of primary creep, and the start of tertiary creep. (There is no real steady state since tests were performed at constant load.) An examination of Fig. 2 reveals a strain at which the strain rate begins to increase rapidly. This strain correlates with that at which cracks greater than the average grain size first appear.

With decreasing stress, final fracture is controlled by the development of a dominant surface crack. Surface cracks grow rapidly with increasing strain at low stresses. Therefore, large surface cracks develop while the level of interior damage is still low. However, the critical crack length for failure appears to be influenced by the amount of internal damage present. As the stress increases, the amount of internal and surface damage become comparable. Near fracture, the level of internal damage is greatest, which suggests that the final fracture is controlled by the development of internal cracking.

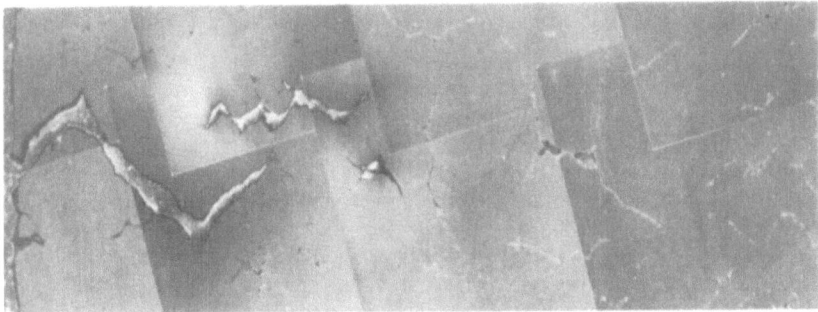

Fig. 12 — Photograph illustrating interaction between a surface crack and interior damage (90X).

It has been suggested (14) that the rate of grain boundary sliding is less stress dependent than deformation by power-law creep. Therefore, at lower stresses, the fraction of strain produced by grain boundary sliding increases and stress concentrations may result at triple points, where sliding is impeded. We find that surface grain offsets and surface cracks initiate early in creep life. At low stresses, these grow rapidly (with strain). This suggests that grain boundary sliding may be important in crack propagation. Because of the extensive stress redistribution which occurs during primary creep, we have attempted to correlate the development of damage with tertiary strain (defined as the strain beyond the strain at minimum creep rate). The onset of tertiary creep is stress dependent (Fig. 3). We find that the tertiary strain correlates damage reasonably well. This is especially true for cracks larger than the average grain size. Therefore, while the growth of small cavities may be stress-dependent, the propagation of wedge cracks is strain-controlled.

The material used in the present study is very similar in chemistry to that used by White et al. (13). However, while our material was sand-cast, theirs was chill-cast, resulting in a more uniform composition and grain size. This led to a difference in both the creep strength and ductility. Furthermore, the accumulation of damage is much more uniform in the chill-cast material. In our material, damage accumulation is rapid, but only in localized regions. This may be due to the higher solute level increasing the matrix creep strength, a grain size effect, or a combination of these which affects the importance of grain boundary sliding.

Once a high-solute creep-brittle region is fully damaged, a crack must propagate through material which is low in solute. The matrix creep strength of these regions is lower, and the stress concentrations due to grain boundary sliding are not as severe. The literature suggests that the rate of creep crack growth varies inversely with matrix strength. In this material, the effect is sufficient to alter the mode of damage, since cavities does not nucleate as easily. The net result is an increase in ductility over material with a uniform solute level of the same average composition. The difference in solute distribution between the two materials may also explain the different creep stress exponents. Inhomogeneous materials often exhibit high creep exponents. An extreme case is eutectic alloys, for which exponents of 20 to 30 are common.

The total average level of damage increases in an approximately linear manner with strain beyond the minimum creep rate. However, there is considerable scatter in the data for a given stress condition. Part of this is due to the approximation of sampling in a finite region in the neck, and using the average strain over the sample area. However, it also appears the fracture cannot be explained solely on the basis of global damage parameters. The distribution of damage is very broad, and the tail end of this distribution is significant in controlling the final fracture process. Observations indicate that the fracture process also involves interactions between the regions of high local damage level and surface cracks (Fig. 12). We therefore need to consider in detail the spatial distribution of damage. Critical local damage levels and/or the spacing between regions of excess damage can be quantitatively analyzed and related to chemical and microstructural variations in the material. This work is currently being completed, and will be reported at a later date.

ACKNOWLEDGEMENTS

The authors would like to express their appreciation to Falconbridge Nickel Co. for supplying the cast material, and to NSERC for financial support in carrying out the experimental investigation.

REFERENCES

1.   Perry, A.J., J. of Mater. Sci. 9 (1974), 1016.
2.   Raj, R., Acta Met. 26 (1978), 995.
3.   Argon, A.S., Recent Advances in Creep and Fracture of Engineering Materials and Structures, B. Wilshire and D.R.J. Owen, eds. (Pinridge Press, 1982), 1.
4.   Cocks, A.C.F. and Ashby, M.F., Prog. in Mater. Sci. 27 (1982), 189.
5.   Svensson, L.E. and Dunlop, G.L., Int. Met. Rev. 26 (1981), 109.
6.   Argon, A.S., Chen, I.W. and Lau, C.W., Creep-Fatigue-Environment Interactions, R.M. Pelloux and N.S. Stoloff, eds. (Met. Soc. of AiME, 1980), 46.
7.   Dyson, B.F., (1982), private communication.
8.   Dyson, B.F., Met. Sci. J. 10 (1976), 349.
9.   Dyson, B.F., Can. Met. Quart. 18 (1979), 31.
10.   Rice, J.R., Acta Met. 29 (1981), 675.
11.   Porter, J.R., Blumenthal, W. and Evans, A.G., Acta Met. 29 (1981), 1899.
12.   White, C.L. and Padgett, R.A., Scipta Met. 16 (1982), 461.
13.   White, C.L., Schneibel, J.H. and Padgett, R.A., Met. Tans. 14A (1983), 595.
14.   Ashby, M.F., Surface Sci. 31 (1972), 498.

SOME FACTORS AFFECTING THE CORROSION FATIGUE PERFORMANCE OF WELDED
JOINTS IN OFFSHORE STRUCTURES

D.J. Burns* and O. Vosikovsky**

*Professor of Mechanical Engineering
University of Waterloo, Ontario

**Research Scientist, Engineering and
Metal Physics Section, Canada
Centre for Mineral and Energy Technology
Ottawa

ABSTRACT

Fixed steel jacket platforms will be used to develop the Venture gas field and semi-submersibles may be used in the development
of the Hibernia oil field. These platforms are fabricated from
large tubular steel sections and a major concern is the fatigue
strength of the welded joints connecting two or more steel tubes.
The wave spectrum causing the fatigue loading is briefly discussed
before considering some of the other environmental factors affecting the corrosion fatigue performance of tubular joints.

1. INTRODUCTION

On February 28th, 1983, Mobil Oil Canada Ltd. and its joint
venture participants issued their environmental impact statement
[1] for the Venture gas field, which is about 210 km off the coast
of Nova Scotia and about 16 km east of Sable Island. The impact
statement says there will be two offshore production complexes and
that each complex will have two wellhead platforms, a processing
platform and an accommodation platform. They propose to use fixed,
steel jacket, platforms at the Venture site, where the water depth
is only 22 to 26 m. This shallow water and the proximity of Sable
Island will have a marked effect on the wave spectrum that the
platforms will experience. Mobil have not as yet made public their

long -term predictions of wave heights at the Venture site. However the impact statement mentions that such platforms are used very successfully in the "southern" North Sea in an environment similar to that at the Venture site. A crude comparison of data published for the Venture area and for the 'southern' North Sea [2] suggests that the comparable North Sea platforms may be some of those operating off Great Yarmouth, U.K., between 53° and 54° N and between 1° and 3° E.

Mobil Oil Canada Ltd. and its partners are expected to submit an environmental impact statement for the Hibernia oil field by November, 1984. This oil field is about 300 km east of St. John's, Newfoundland. The proximity of the Hibernia field to the major drift route for icebergs, and the higher chance of interaction with other types of ice, dictates the types of production system that can be used for Hibernia. Many production systems have been proposed [3], including subsea, but there have been indications that the first system will operate in a water depth of 83 m and use either a large concrete gravity structure or semi-submersibles. Even if semi-submersibles are not used in the first development, they may be used in the development of smaller fields or fields that have to be developed in deeper water. Semi-submersibles will have to be moved to avoid impact with large icebergs, that are not towable, and be designed to withstand impact with some types of ice, such as small bergy bits. At present, drilling semi-submersibles are designed to withstand the impact of a supply boat; for example the DNV rules [4] specify a supply boat of 5000 tonnes displacement and an impact speed of 2 m/sec.

Mobil have not as yet made public any long-term predictions of wave heights at the Hibernia site. A crude comparison of data published by Neu [5] on the 11 year deep-water wave climate for Hibernia with data for the 'middle' North Sea [2] suggests that the wave spectrum at Hibernia may be comparable to that experienced by the North Sea weather station Famita, which operates in water 73 m deep at 57° 30' N, 3° E, or by production platforms, including the semi-submersible BUCHAN, that operate in the region 57° to 60° N of the North Sea.

These crude comparisons between the Venture and Hibernia fields and areas of the North Sea are only intended to give an indication of the severity of the wave loading that East Coast platforms are likely to experience and to emphasize that long established fatigue research programs based on the North Sea environment [6] provide data that are likely to be very useful in the fatigue analysis of East Coast steel platforms.

Figure 1, prepared by Fisher of Lloyds Register of Shipping [7] shows a typical North Sea spectrum and the predicted relationship,

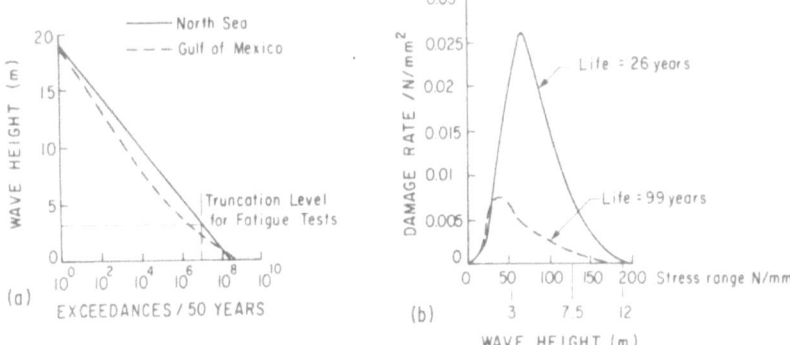

Fig. 1   Wave Climate and Damage Accumulation

Figure 1 (b), between fatigue damage rate and wave height for a
particular structural detail on a fixed platform. Figure 1 also
illustrates the marked difference in wave loading on fixed platforms operating in the North Sea and Gulf of Mexico. Although the
extreme wave heights (50 years) are similar, the spectral differences at more moderate seastates are such that the life estimate
for the North Sea platform is about one quarter that for the Gulf
of Mexico platform. Estimates of extreme wave heights, which have
been published for Venture and Hibernia [1,5], are obviously very
important in limit design, but from a fatigue viewpoint it should
be remembered that Neu [5] has estimated that, for a normal year
at Hibernia, the integrated time for a seastate exceeding 9 m
significant wave height is only three-quarters of a day. Under
more unusual conditions, such as occurred when the Ocean Ranger
sank in February 1982, the nearby Zapata Ugland platform experienced waves exceeding 11 to 12 m for a full day.

The response of a fixed steel jacket platform or semi-submersible to wave loading will obviously be very dependant on detail
design and the orientation of the platform to the principal wave
directions [8]. A major concern is the fatigue strength of the
welded joints, connecting two or more steel tubes [6,9]. The aim
of the remaining sections of this paper is to identify some of the
factors affecting the corrosion fatigue performance of these tubular joints. As mentioned earlier, there are long established
fatigue research programs in Europe [6,10] so most of the fatigue
data mentioned herein was obtained using BS-4360-50 steels or
their European equivalents [11]. These steels have yield
strengths σ_y, of 325-355 MPa and ultimate strengths, σ_u, of 490-
620 MPa.

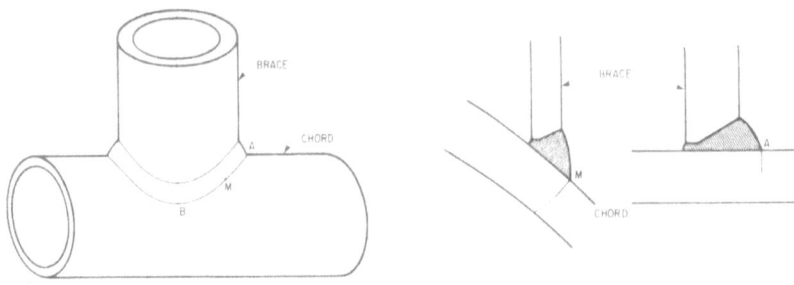

Fig. 2   Hot Spots on T Tubular Joints

Some fatigue crack propagation data for X65 pipeline steel, σ_y of 527 MPa and σ_u of 620 MPa, [12,13] will also be reviewed because it illustrates the influences that hydrogen sulphide pollution and seawater temperature have on fatigue crack growth rate. There has been some research on the combined effect of these two factors on B3-4360-50 steel but the data is not generally available [14].

2.   INITIATION AND PROPAGATION OF FATIGUE CRACKS IN TUBULAR JOINTS

Most of the fatigue data for tubular joints [6,9] has been obtained by testing model joints having the unstiffened T configuration shown in Figure 2. Most of the fatigue tests have been constant amplitude cycling of axial brace load, or in-plane or out-of-plane bending of the brace. Figure 2 shows the critical stress sites or 'hot spots', A, B and M, at which fatigue cracks usually initiate in unstiffened T tubular joints. The cracks usually initiate at the weld toe on the chord tube; for axial brace loading the hot spot is at the saddle point B, but for brace bending, the hot spot location depends on the joint geometry as well as the bending mode, i.e. in-plane or out-of-plane.

Figure 3 and its associated tables, 1(a) and 1(b), show a typical surface crack propagation history for a tubular joint fatigued in air in axial brace tension. This T tubular joint [15] had a chord tube of 914 mm outside diameter and 32 mm wall thickness and a brace tube of 457 mm outside diameter and 16 mm wall thick- ness. Detection of small fatigue cracks at the weld toe of such joints is difficult. In this case, Table 1(a) shows that there was some visual indication at saddle point D after 140,000 cycles and that after 600,000 cycles there was a crack at that location with a surface length of 245 mm. Table 1(b) shows that at saddle point B there was a crack with a surface length of about 60 mm after

Table 1(a)

crack left side of centerline D			crack right side of centerline D		
site of crack tip	length, crack tip to centerline (mm)	total number of cycles	site of crack tip	length crack tip to centerline (mm)	total number of cycles
1	1	140.000			
3	146	600.000	2	99	600.000
5	258	863.000	4	179	863.000
7	273	927.000	6	234	927.000
9	285	980.000	8	299	980.000
11	345	1000.000	10	334	1000.000

A = 140.000

B = 480.000

C = 587.000

D = 600.000

E = 654.000

F = 863.000

G = 927.800

H - 980.000

I = 1000.000

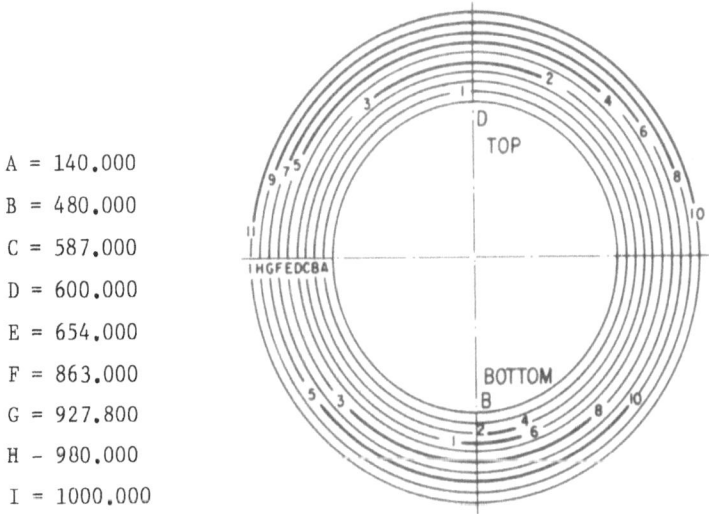

Fig. 3 Surface Crack Propagation History

crack left side of centerline B			crack right side of centerline B		
			2-4	(4)60	480.000
1	25	587.000	6	60	587.000
3	180	654.000	8	160	654.000
5	210	927.000	10	200	927.000

Table 1(b)

58

480,000 cycles. In this test, one of these cracks had propagated
through the chord wall after 950,000 cycles. Tables 1(a) and 1(b)
show that just before through-wall cracking, the surface lengths
of these saddle cracks were 507 mm and 410 mm for D and B respectively.

This crack propagation history and many other examples [15,6]
show that crack initiation occurs after a small fraction of the
life (cycles to through-wall cracking) and that the overall joint
stiffness is not affected significantly prior to through-wall cracking. This crack propagation history also shows that a long surface
crack exists for over 50% of the life.

An indication of the relationship between crack surface length
and depth was obtained by Sanz, Lieurade and Gerald using a crack
marking technique [16]. The data shown in Figure 4 was obtained in
an axial brace tension test on an X-joint. This joint had a chord
tube of 1280 mm outside diameter, with a wall thickness of 77.6 mm,
and brace tubes of 342 mm diameter and 22.4 mm thickness. Cracks
at the saddle points were first observed after 234,000 and 295,000
cycles and through-wall cracking occurred after 1,135,000 cycles.

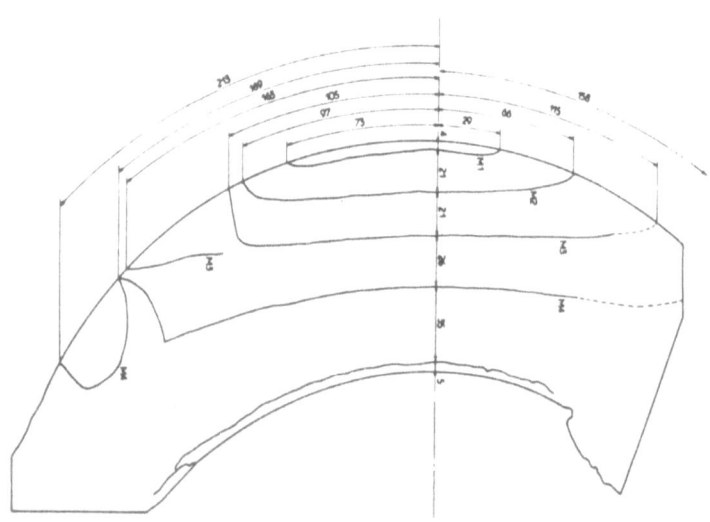

Fig. 4 Propagation of a Fatigue Crack Through the Chord Wall

The crack markings M1, M2, M3 and M4 were made after 330,000, 580,000, 770,000 and 930,000 cycles respectively. The marking M1 shows that after about one quarter of the life, the crack was already 102 mm long and 4 mm deep and that in the next 25% of the life, M1 to M2, the crack grew to be 163 mm long and 25 mm deep.

Crack propagation histories such as Figures 3 and 4 and studies of the early stages of crack growth in tubular joints in air indicate that

1) crack initiation is likely to occur at a small fraction, within 10%, of the number of cyles required to produce a crack through the chord wall.

2) a series of 'semi-elliptic' surface crscks are initiated and coalesce to form a shallow edge crack of the type shown in Figure 4. This coalescense occurs early in the life of the joint eg. M1 on Figure 4; thereafter the rate of growth along the surface drops considerably and the crack grows through the wall. During this phase, the rate of crack growth through the wall is almost constant [23].

This behaviour is difficult to simulate with plate to plate welded specimens, such as the cruciform joints used extensively in the European programs [6], as their cracks tend to be much smaller for a larger percentage of the life and the rate of crack growth accelerates as the depth increases. The situation becomes even more complicated when comparing the behaviour of plate to plate and tubular joints tested in seawater, with or without cathodic protection.

Fisher [17] has pointed out that fixed platforms have a corrosion protection system for the jacket below the waterline and that unprotected joints exposed to seawater are normally those in the splash zone between horizontal jacket levels. He also reports that most semi-submersibles now have corrosion protection, ie., impressed current systems, sacrificial anodes or coatings. In addition one must also consider that such structures are usually painted and that marine fouling may lead locally to the production of hydrogen sulphide.

Very few tubular joints have been tested in seawater, with or without the environmental complications mentioned above; since loading frequency may be important, it is desirable to test at realistic (wave) frequencies, i.e 0.1 to 0.2 Hz. Dijkstra and de Back [18] reported that "the lifetime of their tests (tubular joints) in artificial seawater were about 0.4 times the lifetime of tests in air. There is no difference in the lifetime between the cathodically protected specimen and the unprotected one, but there is a difference in initiation and crack growth". These are important observations that can be best explained by now considering the results of fatigue crack propagation studies on plate specimens in seawater.

3. FATIGUE CRACK PROPAGATION STUDIES IN SEAWATER

Extensive research programs on quantifying the effects of loading and environmental parameters on fatigue crack growth rate in B44 360-50D steel, or equivalent, in seawater have been underway for some years in Europe. The effects of cycle frequency, stress ratio, electrochemical potential, intermittent immersion and oxygen content have been investigated [19]. These tests were conducted at temperatures which prevail in the North Sea (5 to 12°C), or at room temperature. There is some evidence that the colder sea water is less detrimental to fatigue life. Since the temperature of most Canadian offshore waters is lower than that used in the European studies, a study was made at CANMET [12], using X65 steel, of fatigue crack propagation rates in artificial seawater at 0°C and room temperature at two electrochemcial conditions, free corrosion and cathodically protected at -1.04 V.

Fatigue crack growth rates were measured on single-edged notched, pin-loaded specimens 13.4 mm thick and 76.2 mm wide. The specimens, with T-L orientation, were cut from a X65 pipeline steel plate, manufactured to API 5LX specification.

Tests were carried out under constant amplitude sine-wave loading in a computer-controlled 500 kN MTS system. Prior to testing in sea water, the specimens were pre-cracked in air at high frequency using a load shedding procedure. After pre-cracking, the crack extended at least 2.5 mm from the tip of the initial notch, and the stress intensity range, ΔK, was between 10 and 11 MPa $\sqrt{m}$ for tests at R = 0.05 and between 9 and 10 MPa $\sqrt{m}$ for tests at R = 0.5. Crack length, a, was measured simultaneously by the DC potential drop technique and by two travelling microscopes. Crack length measurements made by the potential drop technique had an estimated accuracy of 0.05 mm. Microscope measurements were used to calibrate and check potential drop measurements.

Artificial sea water, prepared from distilled water according to ASTM D1141-75, was kept in an insulated tank, either at room temperature or cooled to -2 to 0°C. During the tests, the water was circulated through a lucite chamber, mounted on the specimen over the notch, at a rate of 1 L/min. Air was continuously bubbled through the water in the main tank. The pH was maintained at 8.1 $\pm$ 0.2. A new batch of water was mixed after every second test.

The temperature of the water in the test chamber varied from -1 to +1°C during low temperature testing and from 23 to 27°C during room temperature testing. The electrochemical potential was measured periodically using a saturated calomel electrode. In the free corrosion tests, the potential varied from -0.66 to -0.71 V. A cathodic potential of -1.04 ± 0.02 V was achieved by coupling the specimen to two cylindrical zinc anodes.

After pre-cracking in air, the test specimen was immersed in sea water and the crack extended by about another 0.5 mm at high frequency. The frequency was then reduced to the test frequency of 0.1 Hz. To accelerate the test, the loads were progressively raised by 5%. Initially, the load was increased after one day of testing (8600 cycles). As the crack grew faster, the load was increased after an increment of 0.3 to 0.5 mm. Towards the end of the test, the increment was raised to 1 mm and the loads were kept constant. The initial (one day) crack growth increments were small (0.1-0.3 mm) and as a result, the growth rates below 10^{-5} mm/cycle are subject to greater error. At cathodic potentials, some initial crack growth was noted, but after about one day of cycling, crack growth arrested. The crack remained dormant for several days (several load increases) before starting to grow again. The duration of a test was between two and three weeks. Growth rates were calculated using the secant method - ASTM E647-81.

The dependence of fatigue-crack-growth rates, da/dN, at 25 and 0°C, on the stress intensity factor range, ΔK, is shown in Fig. 5-8. Figures 5 and 6 show growth rates in X65 steel at free

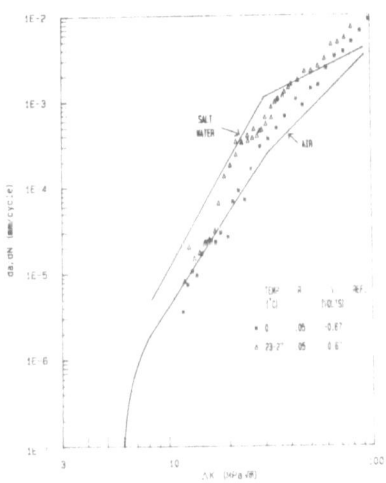
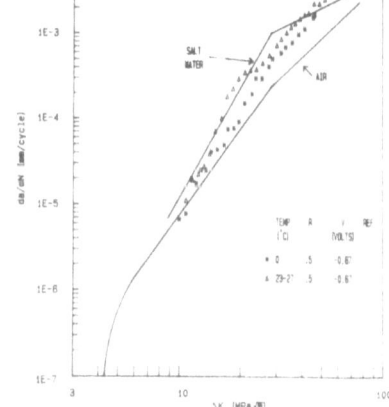

Fig. 5 Crack Growth Rates
R = 0.05, f = 0.1 Hz
T = 0 and 25°C.

Fig. 6 Crack Growth Rates
R = 0.5, f = 0.1 Hz
T = 0 and 25°C.

corrosion potential for stress ratios R = 0.05 and 0.5 respectively.
The corresponding data for cathodic potential of -1.04 V, are
shown for X65 in Fig. 7 and 8. The solid lines in these figures
are mean growth-rate curves, reported previously, for pipeline
steels at room temperature in air and 3.5% NaCl solution (20.21).
For low-stress ratios (Fig. 5 and 7), these mean curves are for
X65 steel; for high-stress ratios (Fig. 6 and 8), these mean curves
are for X70 steel at R = 0.4. Since X65 and X70 have similar
growth rates, in air or salt water at low R values, it is believed
that at high R Values, X70 and X65 should have essentially identical growth rate characteristics.

3.1 Free Corrosion Data

At the free corrosion potential (Fig. 5 and 6), room temperature growth rates in sea water are comparable with those in salt
water. In the intermediate ΔK range, the growth rates in both
solutions are about four times higher than those in air.

A drop in sea water temperature to 0°C results in a reduction
of the intermediate ΔK range growth rate by a factor of two at
R = 0.05 and somewhat less at R = 0.5. Consequently intermediate
ΔK, fatigue-crack growth rates in 0°C sea water under free corrosion are still somewhat higher than growth rates in room-tempera-
ture air.

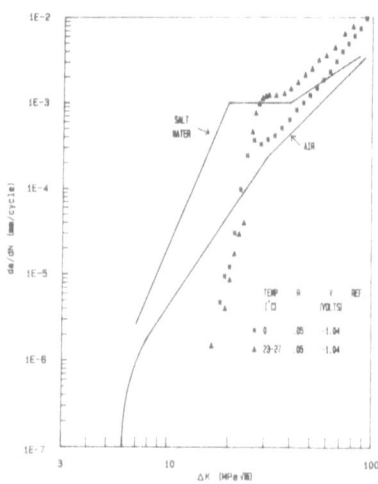
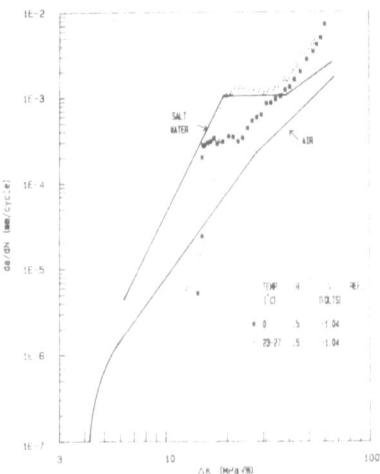

Fig. 7 Crack Growth Rates at
        -1.04V R = 0.05, f =
        0.1 Hz, T = 0 and 25°C.

Fig. 8 Crack Growth Rates at
        -1.04V R = 0.5, f =
        0.1 Hz, T = 0 and 25°C.

3.2 Cathodic Protection Data

At a cathodic potential, (Fig. 7 and 8) the pattern of environmental growth-rate enhancement is more complex. The growth-rate curves in salt water exhibit a characteristic plateau where the growth rate is independent of ΔK in the range about 20 to 40 MPa $\sqrt{m}$. The maximum environmental enhancement of growth rate, about one order of magnitude in salt water at room temperature, occurs at the beginning of the plateau, where $\Delta K \sim 20$ MPa $\sqrt{m}$. The sea water growth rates at room temperature exhibit a plateau at the same da/dN value as in salt water with the same upper ΔK limit of ~40 MPa $\sqrt{m}$. However, the full extent of the plateau in sea water is preserved only at a stress ratio R = 0.5 (Fig. 8). At a stress ratio of R = 0.05 (Fig. 7), the plateau is not very distinct.

Reduction in seawater temperature from room to 0°C decreased the plateau growth rates by almost a factor of four. Again the plateau is rather indistinct at the lower stress ratio.

Below these plateaux, the sea-water growth rates drop much below those in air. Extrapolation of the growth-rate curves give apparent threshold stress intensity ranges of ~15 MPa $\sqrt{m}$ for R = 0.05, and of ~10 MPa $\sqrt{m}$ for R = 0.5; almost three times as high as the thresholds in air or salt water. The suppression of crack growth at low ΔK values is believed to be caused by precipitation of calcareous deposits within the crack [22]. These deposits, which are a product of cathodic reaction, wedge the crack open and thus reduce the effective stress-intensity range. If the calcareous deposition rate is high enough in relation to the growth rate and crack opening, the deposits can reduce the effective stress-intensity range below the threshold level and crack growth stops. At higher ΔK, the crack outgrows the deposition rate, and the growth rate reaches that in salt water.

3.3 General Discussion of X65 and Other Fatigue Crack Propagation Data

The data presented in Figures 5-8 for X65 steel and data obtained by Scott and Silvester for BS 4360-50 steel show that, at intermediate ranges of ΔK, there is a significant reduction in fatigue crack growth rate as the seawater temperature is reduced from 25° to 0°C.

At intermediate ranges of ΔK, the growth rates in X65 in seawater are higher than those in air and the effects of cathodic protection are consistent with those observed by Scott and others. For specimen crack geometrics and growth rates where calcareous blocking is insignificant, the crack growth rates increase when cathodic protection is used. Figure 9, prepared by Walker [19],

64

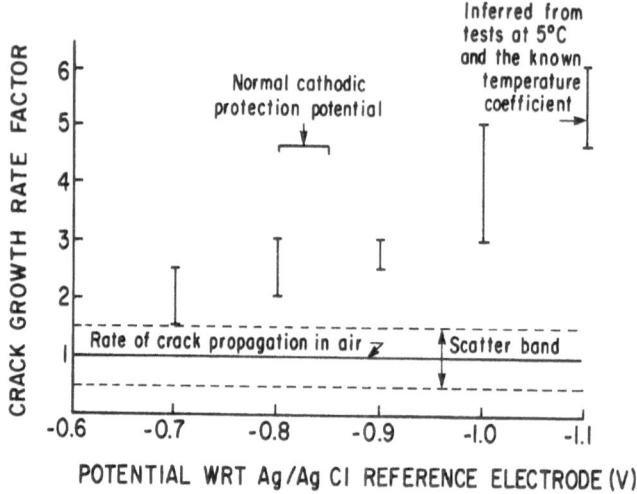

Fig. 9   Relative Fatigue Crack Growth
Rates in Air or in Seawater.

shows relative crack growth rates for BS-4360-50 steels and their
equivalents tested in air or in seawater at various potentials;
the free corrosion potential in these tests was about 0.65 to 0.7V.
It seems to be generally agreed that this increase in crack growth
rate with increase in cathodic protection is the result of an
increase in hydrogen uptake at the crack tip, which leads to a more
severe embrittlement of the crack tip plastic zone.

These observations on hydrogen embrittlement and earlier observations on the effect of calcareous blocking on growth rates at
lower ranges of ΔK can be used to explain the observations made by
Dijkstra and de Back [18] during their tubular joint tests in seawater. As mentioned earlier, cracks initiate at an early stage in
the life of a tubular joint. While the cracks in tubular joints
are very small, one would expect calcareous blocking to be most
effective. In other words, cathodic protection should tend to
slow the growth of small cracks, but as the crack increases in
size, the blocking should be less effective and the increased
hydrogen embrittlement produced by cathodic protection should lead
to crack growth rates higher than those observed in tubular joint
without cathodic protection. Apparently the balance between these
competing effects, calcareous deposition and hydrogen embrittlement, in Dijkstra's tests was such that overall there was no difference in total life although the crack "incubation" period was
longer with cathodic protection.

These observations also have some important implications when interpreting endurance tests on small plate to plate welded joints in seawater, with and without cathodic protection. If the joint geometry and method of loading are such that the fatigue cracks tend to remain small for most of the fatigue life, then calcareous blocking may be much more significant than it is in tubular joint tests. In other words, _endurance_ tests on small plate to plate joints may not be a good indicator of the effects that seawater and cathodic protection have on the endurance of large tubular joints.

As mentioned earlier, another effect that must be considered is that marine fouling may lead locally to the production of hydrogen sulphide. Some indication of the effects of this can be obtained by considering crack growth data obtained by one of the authors [13] for X65 steel in a sour oil environment. In crude oil saturated with H_2S (5000 ppm) growth rates were increased by an order of magnitude compared to those in air. As the H_2S content was decreased to 30 ppm, there was a marked reduction in its effect on crack growth rate. However it is far from clear what the combined effects would be for a cathodically protected tubular joint, immersed in seawater, with local H_2S pollution.

Finally, it should be mentioned that much of the fatigue data for tubular joints and fatigue crack growth data for joint steels has been obtained in constant amplitude tests. Spectrum fatigue tests designed to simulate wave loading, Fig. 1(a), including the marked differences between summer and winter seastates, might well alter the balance between the factors influencing crack growth rate and modify some of the conclusions drawn from constant amplitude data.

The number of waves seen by an offshore structure exceeds 10^8. To keep test times within reasonable bounds, particularly at realistic frequencies, 0.1 to 0.2 Hz, one must truncate the load spectrum as shown schematically in Figure 1(a). The choice of this truncation level and the return period for spectrum tests have been the cause of much debate [24] and only now have the European laboratories come to a tentative agreement on a common load spectrum for part of the next phase of the ECSC program.

66

REFERENCES

1. Mobil Oil Canada Limited, Venture Development Project Environmental Impact Statement, Vol. 1, Summary February 1983.
2. Burns, D.J., Survey of Welded Joint Specifications for Eastern
   and Arctic Offshore Structures for Physical Metallurgy Research
   Laboratories, CANMET July, 1983.
3. Petroleum Directorate, Govt. of Newfoundland and Labrador,
   Proceedings of Symposium on Production and Transportation
   Systems for the Hibernia Discovery, February 1981.
4. Det Norske Veritas. Rules for classification of mobile offshore units March, 1981.
5. Neu, H.J.A., Eleven Year Deep-water Wave Climate of Canadian
   Atlantic Waters, Canadian Technical Report of Hydrography and
   Ocean Sciences No. 13, Department of Fisheries and Oceans,
   October 1982.
6. Commission of the European Communities, Proc. of International
   Conference on Steels in Marine Structures, Paris, October, 1981.
7. Fisher, P.J., Proc. of Conf. on Fatigue in offshore structural
   steel, ICE London, 1981, Session 7, Summary of Current Design
   and Formulation of Guidance Notes, p. 93-101.
8. Bainbridge, C.A., Design and Operation of Semi-submersibles.
   Presented at the Royal Institution of Naval Architects Symposium on Offshore Engineering, Nov. 1981, LLoyds Register of
   Shipping report No. 78.
9. Rodabaugh, E.C. Review of Data Relevant to the design of
   tubular joints for use in fixed offshore platforms. Welding
   Research Council Bulletin 256 January 1980.
10. Department of Energy, U.K. The United Kingdom Offshore Steels
    Research Projects (UKOSRP 1 and 11).
11. Harrison, J.D. Material Selection Considerations for Offshore
    Steel Structures, Ibid (6), Special and Plenary Sessions
    Volume, p. 147-193.
12. Vosikovsky, O., Neill, W.R., Carlyle, D.A. and Rivard, A.. The
    Effect of Seawater Temperaature on Corrosion Fatigue Crack
    Growth in Structural Steels. Physical Metallurgy Research
    Laboratories Report ERP/PMRL 83-27 (OP-J), April, 1983.
13. Vosikovsky, O., Corrosion Fatigue Life of a Sour Crude Oil
    Pipeline. Physical Metallurgy Research Laboratories Report
    MRP/PMRL 82-7 (OP-J), January, 1982.

14. Webster, S.E., Austen, I.M. and Rudd, W.J., British Steel
    Corporation, U.K., Confidential Communication, April, 1984.
15. de Back, J. and Vaessen, G.H.G., Fatigue and Corrosion Fatigue
    Behaviour of Offshore Steel Structures, Final Report ECSC Con-
    vention 7210-KB/6/602 (J71f/76) April 1981.
16. Sanz, G., Lieurade H.P. and Gerald J., Fatigue Tests on Ten
    Full-Scale Tubular Joints Ibid 6, Technical Session 8, paper
    8.1.

17. Fisher, P.J., Revised fatigue guidance and the certifying authority. Welding Institute Conference on Offshore Structures, U.K., November, 1982, paper 57.
18. Dijkstra, O.D. and de Back, J., Fatigue Strength of Tubular T and X-joints, Offshore Technology Conference (12th) Houston, May, 1980 OTC paper 3696.
19. Walker, E.F., Effect of Marine Environment Ibid 6, Special and Plenary Sessions Volume, p. 195-251.
20. Vosikovsky, O., Journal of Testing and Evaluation, J. TEVA 8:2, March, 1980, p. 68-73.
21. Vosikovsky, O., Journal of Engineering Materials and Technology, Trans. ASME 97:4, Oct. 1975, p. 298-302.
22. Scott, P.M., Thorpe, T.W. and Silvester, D.R.V., Rate determining processes for corrosion fatigue crack growth in ferritic steels in seawater, European Federation of Corrosion, Meeting on Low Frequency Cyclic Loading Effects in Environmental Sensitive Fracture, Milan, March 9-11, 1982.
23. Dover, W.D. and Dharmavasan, S., Fatigue Fracture Mechanics Analysis of T and Y Joints, Offshore Technology Conference, Houston, 1982, p. 315-326.
24. Schutz, W., Procedures for the Prediction of Fatigue Life of Tubular Joints. Ibid (6) Special and Planary Sessions Volume, p. 253-308.

THE FRACTOGRAPHY OF ENVIRONMENTALLY ASSISTED CRACKING

J.I. DICKSON and J.P. BAÏLON

ECOLE POLYTECHNIQUE
Montreal, Quebec, Canada, H3C 3A7

ABSTRACT

Some of the fractographic aspects of environmentally assisted cracking are reviewed, with particular emphasis on the fractography of transgranular stress corrosion cracking and of corrosion-fatigue in dual or multiphase materials. The strong similarity between the fractography of transgranular stress corrosion cracking, cleavage produced by hydrogen embrittlement, cyclic cleavage and cracking produced near the fatigue threshold indicates that all these types of cracking occur by a discontinuous cleavage mechanism. Corrosion-fatigue or stress corrosion cracking can give rise in engineering materials to a microscopic crack front that leads the macroscopic front in one phase or along a microstructural feature and which results in considerably accelerated crack growth.

1. INTRODUCTION

An important type of time-dependent fracture is environmentally assisted cracking. In studying crack propagation by stress corrosion cracking (SCC), corrosion-fatigue or hydrogen embrittlement (HE), fractographic observations should be particularly useful in identifying or verifying the cracking or accelerated cracking mechanisms. The objective of the present paper is to review and discuss some of the fractographic features associated with environmentally assisted cracking and where appropriate to compare these features with those produced by other types of cracking.

2. FRACTOGRAPHIC FEATURES

2.1 Fractography of SCC

Either a transgranular or an intergranular cracking mode can occur during SCC, with at times a transition in mode with stress intensity factor, K (1-3), electrochemical potential (1) or degree of cold-working (1, 2, 4). Intergranular SCC often appears consistent with cracking produced by film rupture and anodic dissolution (4); however, HE, liquid metal embrittlement and for certain conditions, cracking in the absence of environmental effects can also occur intergranularly. Segregated solutes (5) or precipitates (6) at grain boundaries, moreover, can favour environmentally-assisted intergranular cracking.

The fractography of transgranular stress corrosion cracking (TSCC) has been quite extensively studied (2, 4, 7-11). This fractography is generally highly crystallographic and appears quite similar in different metals of different crystalline structure. The fine microscopic details on opposite fracture facets generally appear to interlock perfectly. Features which appear to be crystallographically serrated river lines are present on the fracture facets. These multiply on crossing grain boundaries and can be employed to follow the crack propagation from grain to grain. These features generally suggest that cracking occurs by a cleavage-type mechanism. Cracking occurring by a series of small discontinuous crack jumps is supported in particular by the work of Pugh and co-workers (4, 7). Most of the fractographic features, however, are not completely inconsistent with the occurrence of very fine crystallographic dissolution, as indicated by the high-voltage transmission electron microscopy observations of Scamans and Swann (11).

Very crystallographic striation-like markings, roughly perpendicular to the river lines (Figure 1), are observed (2, 4, 9, 10) and have been studied in 304 and 310 austenitic stainless steels (9, 10). In these metals, the crack propagation occurs on {100} planes in a <110> direction, with the striations parallel to the second <110> direction contained in the crack plane. This is also the direction in which slip traces formed at the crack front would be found. These surface steps corresponding to the striations interlock (2, 4, 10) on opposite fracture surfaces and if they correspond to slip traces, the slip must have occurred exactly at the crack tip for this transgranular cracking. The striations become more marked as K increases (8, 10) but the interstriation spacing appears to be much less strongly dependent on K than the crack velocity (9, 10). With this cracking crystallography, a difficulty exists in distinguishing between striations which correspond to temporary crack arrest sites and those which only correspond to slip traces. At

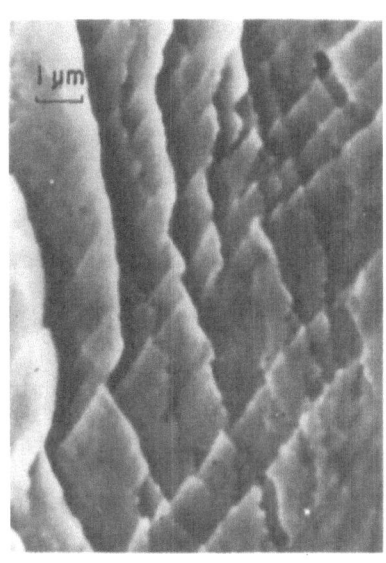

Fig.1 Jogs on river lines to follow SCC striations; 310 steel in boiling MgCl$_2$ solution (10).

Fig.2 SCC striations and river lines on 310 steel tested in boiling MgCl$_2$ solution (2).

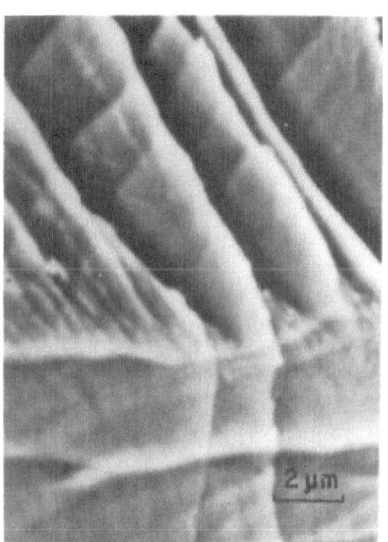

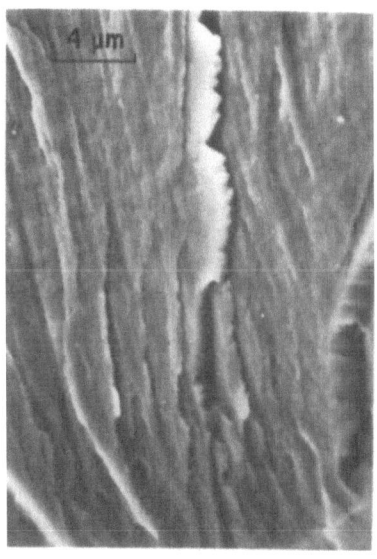

Fig.3 Cyclic cleavage striations on a renitrogenized mild steel (12).

Fig.4 SCC striations and serrations on river lines on 310 steel (10).

times river lines can be seen to start or become more pronounced at striations (8, 10) or to change directions to follow striations for a short distance (Figure 1). Such features were also found for cyclic cleavage striations in a renitrogenized mild steel (12) fractured in the absence of environmental effects. This strongly suggests that at least some of the striations correspond to true crack arrest sites. The strong similarity between TSCC and cyclic cleavage can also be seen by comparing Figures 2 and 3.

The frequent presence of crystallographically serrated river lines (Figure 4) can be explained (4,7) by the occurrence of macroscopically slow cleavage at low stresses which favours the fracture of the ligaments between neighbouring parallel cracks by low energy mechanisms. In support of this explanation, it can be noted that serrated river lines have also been observed associated with cyclic cleavage at -40^{0}C in a mild steel (13) as well as for HE (14, 16). The similarity between the TSCC fractographic features and those produced by cathodic charging in 316 stainless steel (15) and in b.c.c. Fe and Fe-Si (16) further suggests that this fractography is produced by small discontinuous cleavage-type jumps.

For TSCC of h.c.p. zirconium and titanium alloys, flutes can be found which do not interlock on opposite surfaces and which correspond to ductile fracture permitting parallel crystallographic cleavage-type cracks to join together (17).

For intergranular SCC, reliable identification of striation markings as true crack arrest sites is rare. The striations observed by Vermilyea (5) on 304 steel and on Inconel 600 tested in hot sulfate solutions appear to have the characteristics of true crack arrest markings. In general, slip lines (Figure 5) which interlock on opposite fracture surfaces can be found (13) but on intergranular facets these may have been produced during prestraining or ahead of the crack tip. For 316 stainless tested in boiling MgCl$_2$ solution, these were observed to become more marked as K increased (18) and occasionally transgranular cracks (Figure 5) could be seen to initiate along such slip traces. At times, a line of crystallographic pits along a slip trace were found on one intergranular facet and in such cases, a line of interlocking protrusions (Figure 5) on the opposite surface could be found, showing that if dissolution directly causes the initiation of this transgranular cracking, it is extremely fine. On aluminium alloys (19), sub-boundaries can be found on intergranular facets.

2.2 Similarity Between TSCC and Fatigue Threshold Fractography

As pointed out previously (20-23), a striking similarity exists between the fractography of TSCC and that produced by transgranular cracking near the fatigue threshold including in inert environments. This similarity has also been explained by near-threshold fatigue cracking occurring by a series of small, discontinuous cleavage-type crack jumps (22, 23). A discontinuous cracking mode is required near the fatigue threshold where the average crack rate can be less than one interatomic spacing per cycle. The other explanation often proposed (24) is that near-threshold propagation corresponds to slip plane cracking. The crystallography of the fractographic features produced at the threshold in cubic metals (20, 23, 25) is often {100} planes with the river lines indicating <110> or < 100 > propagation directions. Facets corresponding to {110} or {111} planes have also been reported (18, 25). Figure 6 shows an example of a {111} facet observed in the α-phase of a 12% Mn-Ni-Al bronze. Nevertheless, in this material, most near-threshold facets appeared to correspond to {100} planes (26). It can be concluded that generally the near-threshold facets do not correspond to slip planes. When {111} facets are present these can often be easily recognized and may be reported in disproportionate frequency to their occurrence. Even though similar facets are also produced for near-threshold fatigue propagation in such inert environments as dry argon (23, 27), silicone oil (23, 24) and vacuum (28), it is difficult to completely rule out a possible role of the environment, since the crack growth rates involved are so slow that time is available for the crack tip material to react with even trace amounts of aggressive species such as water or water vapour. This possibility is further suggested by the frequent occurrence of some intergranular cracking at somewhat higher crack growth rates (20, 23, 27). Marchand (27) studied near-threshold fatigue crack propagaion in copper and 70-30 α-brass of two grain sizes in air and in dry argon and found that transgranular to partially intergranular to transgranular fractographic transitions observed as a function of crack growth rate were consistent with an environmental effect. It can be suggested that a detailed study of the influence of cycling frequency on these fractographic transitions, preferably performed at a high R-ratio so that crack closure effects which are particularly important in the near-threshold region will not influence the results, could be useful in verifying a possible role of the environment in such cracking.

2.3 The Fractography of Corrosion-Fatigue

The occurrence of corrosion-fatigue crack propagation is often indicated by a change from ductile to brittle striations (29-32). Other than that the brittle striations are generally

74

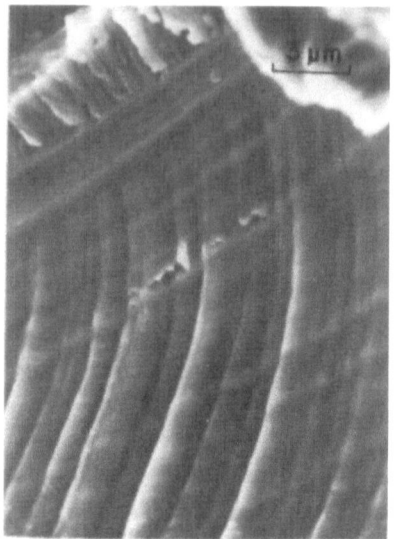

Fig.5 TSCC and protrusions
nucleated on intergranular slip
traces in 316 steel (18).

Fig.6 Fatigue near-threshold
{111} facet on 12% Mn-Ni-Al
bronze tested in air (26).

Fig.7 Crystallographic cracking
with striations produced by high
temperature fatigue of Inconel
X-750.

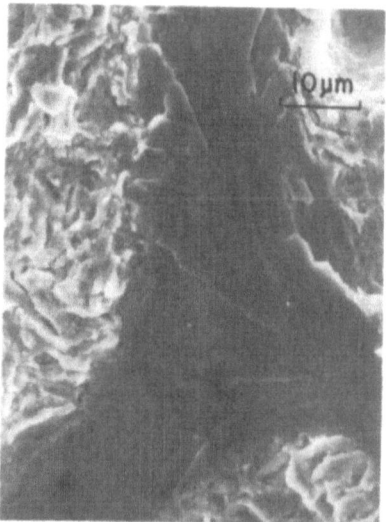

Fig.8 Brittle striations in
ferrite and ductile striations
in martensite for corrosion-
fatigue of CA 15 (32).

more marked than those produced by TSCC, the fractographic features of corrosion-fatigue often resemble closely those of TSCC. Similar fractographic features (Figure 7) can also be obtained for high temperature fatigue cracking, where an environmental effect can also cause accelerated crack growth (33).

In dual or multiphase materials, the different phases can have different susceptibilities to environmentally assisted cracking. An example (30, 31) was observed during the corrosion-fatigue of cast KCR 171, a duplex stainless steel whose microstructure consists of small islands of austenite occupying approximately 55% of the volume within large ferrite grains. When tested in white water (a pulp and paper industry environment), the ferritic phase was subject to corrosion-fatigue for certain temperatures and cycling frequencies. The accelerated crack growth in the ferrite permitted the austenite grains to be completely bypassed by the crack front in the ferrite. The uncracked austenite grains behind this front then fractured much in the same manner as low cycle fatigue samples from multi-initiation sites at the ferrite-austenite interface. In comparison for testing in air at 20^0C, both phases showed ductile striations with similar interstriation spacings. The occasional ferritic grain, however, still presented brittle striations showing that the occurrence of some corrosion-fatigue could occur in this environment, a result which can be explained by the reaction of fresh crack tip material with water vapour to produce oxide and hydrogen (34).

A more striking example of corrosion-fatigue effects in a dual-phase material was observed for CA-15 stainless steel (32), containing approximately 95% austenite and 5% ferrite, tested in white water. Again the ferritic phase was more susceptible to corrosion-fatigue and at higher ΔK, the martensitic phase presented ductile striations (Figure 8). Nevertheless, the accelerated crack growth in the ferrite caused the microscopic crack front to proceed ahead of the macroscopic front at numerous sites, with the uncracked ligaments between such sites then fracturing partly by crack propagation in a direction almost perpendicular to the macroscopic crack front. The resulting acceleration of the macroscopic crack rate was comparable to that observed for KCR 171.

Two other examples of narrow microscopic crack fronts proceeding ahead of the macroscopic front at numerous sites and resulting in a significant acceleration of the crack rate can be cited. The first concerns the obtention of corrosion-fatigue behaviour in a 12% Mn-Ni-Al bronze in 3.5% NaCl solution. In this case, corrosion-fatigue resulted (35) in very narrow slotting at α-β boundaries proceeding ahead of the macroscopic crack front and nucleating ductile striations at different sites in

α-grains, which under certain conditions could even become essentially surrounded by these slots (Figure 9). The second
example concerns the influence of cold-rolling on SCC velocities
on 316 and 310 austenitic steels (1, 2, 18) tested in boiling
$MgCl_2$ solution with the crack propagation plane perpendicular to
the (long) transverse direction and the cracking direction parallel to the rolling direction. For the cold-rolled 316 stainless steel, narrow microscopic intergranular cracking largely
perpendicular to the short transverse direction proceeded ahead
of the macroscopic crack front at sufficiently high K values,
with the transgranular cracking in the ligaments between these
intergranular cracks then proceeding almost at 90^0 to the macroscopic direction. The occurrence of this effect in cold-
rolled 316 steel and its absence in annealed 316 and 310 and in
cold-rolled 310 steels appears to explain (18) the accelerated
crack growth caused by cold-rolling the 316 steel and the absence of such an influence for the 310 steel.

In contrast, Dickson et al (13) studied the influence of
the occurrence of clusters of cleavage facets on the fatigue
crack growth of a mild steel at low temperatures. This cleavage
caused the microscopic crack front to proceed ahead of the macroscopic front. For certain conditions, up to 50% of the fracture surface could be covered with cleavage facets before a
significant acceleration in macroscopic growth rate resulted.
In this case, the lack of influence on the macroscopic growth
rate of the leading microscopic crack fronts appears related to
the relatively large ligaments, covered with ductile striations
on the fracture surface, between cleavage clusters.

The examples cited show different influences that a semi-
cohesive crack tip zone can have on the cracking kinetics. The
indication is that, in environmentally assisted cracking, even
very narrow microscopic cracking proceeding ahead of the macroscopic crack front at numerous finely-spaced sites can result in
considerably accelerated cracking.

2.4 Fractography of Hydrogen Embrittlement

Hydrogen can give rise to cleavage or quasi-cleavage in
ferrite or in tempered martensite. As mentioned, the aspects of
this cleavage can be similar to that observed for cyclic cleavage (12, 13). A number of authors, however, have reported
{110} or {112} facets (15,36) for HE and have proposed cracking
by slip plane decohesion. During cyclic cleavage of a renitrogenized mild steel in ordinary air (12), very similar facets
were observed (Figure 10 and the lower grain of Figure 3), when
the amount of plasticity accompanying this cleavage was particularly important. Further work appears required to determine
clearly whether such facets are produced by slip or by a clea-

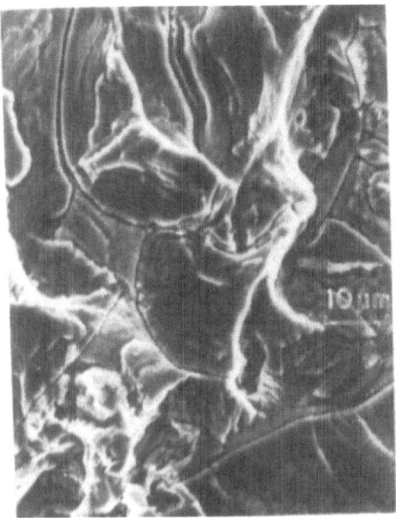

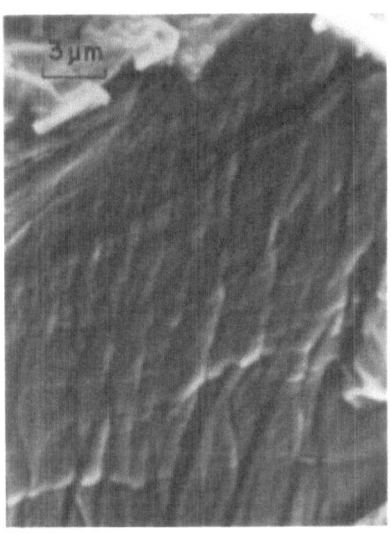

Fig.9 Corrosion-fatigue of Mn-Ni-Al bronze with ductile stria-tions nucleating at slots at α-β interfaces (35).

Fig.10 Semi-brittle cyclic cleavage produced in assumed absence of environmental ef-fects at 100°C (12).

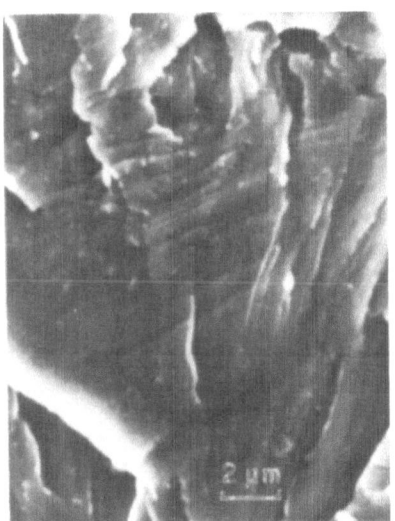

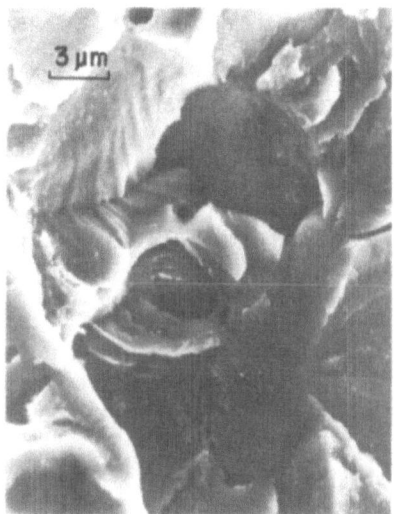

Fig.11 Cleavage striations pro-duced by hydrogen - assisted fracture of A-36 steel (16).

Fig.12 Intergranular stria-tions produced by hydrogen-assisted fracture of A-36 steel (16).

vage mechanism. The occurrence of {110} facets in tempered martensite can be explained by decohesion along martensite lath interfaces (36). In 304L austenitic stainless steel, Caskey (37) observed that hydrogen charging caused parting along {111} interfaces of annealing twins.

Figures 11 and 12 present fractographs from a multiple initiation slow-cracking region of a galvanized bar of A36 ferritic-perlitic mild steel. Strong evidence existed in this failure analysis (16) that this slow cracking was hydrogen assisted. The cleavage in the slow cracking regions covered by blackish corrosion product prior to cleaning presented crystallographic details differing from the cleavage in the fast fracture regions. In the former regions, a small fraction of intergranular facets at times presenting striation-like markings were found. As well, striations or other crack arrest marks were found on some of the transgranular cleavage facets and frequent reinitiation occurred ahead of the macroscopic crack front either at grain boundaries or at times at intragranular particles with river lines, which in some cases contained crystallographic serrations, radiating away in all directions. Similar initiation at inclusions has been reported for transgranular cracking of cathodically charged 4130 steel (6).

Beachem (3) for AISI 4340 martensitic steel showed transitions from intergranular to quasi-cleavage to ductile striations as K increased for both HE and SCC. Gangloff and Wei (38) in a Maraging steel did not observe this transition as a function of K but observed a transition as a function of test temperature. The present authors for service failures of different high strength steel bolts have generally observed a transition from intergranular to ductile tearing simples as a function of K with the intermediate quasi-cleavage region absent or only a few grains in width, but were not able to identify clearly the transition between HE and overload fracture, since it is known (39) that hydrogen can influence both the nucleation and the growth of microvoids during ductile fracture.

In the case of through-thickness crack propagation in laboratory samples, a difference often can be expected for hydrogen-assisted cracking between the fractographic features near the sides where plane stress conditions prevail and those further in the interior where plane strain conditions prevail. Two such examples will be cited associated with corrosion-fatigue behaviour identified as caused by HE. In the case of fatigue propagation in KCR 171 stainless steel (30, 31) in white water, brittle striations were observed in the ferritic phase except near the lateral edges of the fracture surface, where a region of ductile striations, which increased in width with increasing ΔK, was present. Similar effects were found for SAE 304 and for the ferritic phase of CA 15 (32), although for these two steels occasional groups of brittle striations could also be found near the lateral edge. For crack propagation of a martensitic steel in lubricating oil containing traces of dissolved water, Delebecq et al (40) found corrosion-fatigue effects with intergranular cracking except near the lateral surfaces.

On the other hand, in the case of TSCC of such f.c.c. metals as austenitic stainless steels and Admiralty metal where hydrogen embrittlement is increasingly considered as an attractive explanation, there does not appear to be a significant difference between the fractographic features near the sides and in the interior (2, 4, 7, 18). The other fractographic aspects, as pointed out, are consistent with a discontinuous cleavage mechanism. As well, even in cases of clear HE, a difference in fractography at the edges is not always observed (39).

3. CONCLUSIONS

From the fractographic observations reviewed, it can be concluded that considerable information can be obtained from detailed fractographic studies of environmentally assisted cracking. Such detailed fractography is often underutilized in such studies, despite the requirement that the proposed crack propagation mechanisms must be compatible with the fractographic features.

4. ACKNOWLEDGMENTS

The authors are indebted to their coworkers cited in the references for the permission to employ unpublished results and observations. Financial assistance from the FCAC (Qubec) and NSERC (Canada) programs is gratefully acknowledged.

REFERENCES

1. Russell, A.J. and Tromans, D. A Fracture Mechanics Study of Stress-Corrosion Cracking in Type-316 Austenitic Steel. Metallurgical Transactions 10A (1979) 1229-1238.
2. Dickson, J.I., Russell, A.J. and Tromans, D. Stress Corrosion Crack Propagation in Annealed and Cold Worked 310 and 316 Austenitic Stainless Steels in Boiling (154^{0}C) Aqueous MgCl$_2$ Solution. Canadian Metallurgical Quarterly 19 (1980) 161-167.
3. Beachem, C.D. A New Model for Hydrogen-Assisted Cracking (Hydrogen "Embrittlement"). Metallurgical Transactions 3 (1972) 437-451.

4.   Bursle, A.J. and Pugh, E.N. An Evaluation of Current Model
     for the Propagation of Stress-Corrosion Cracks. Environ-
     ment-Sensitive Fracture of Engineering Materials (New York,
     AIME, 1979) pp 18-47.

5.   Vermilyea, D.A. Stress-Corrosion Cracking of Iron and Nic-
     kel Base Alloys in Sulfate Solutions at 289°C. Corrosion 29
     (1973) 442-448.

6.   Craig, B.D. and Krauss, G. The Structure of Tempered Mar-
     tensite and Its Susceptibility to Hydrogen Stress Cracking.
     Metallurgical Transactions 11A (1980) 1799-1808.

7.   Bursle, A.J. and Pugh, E.N. On The Mechanism of Transgra-
     nular Stress-Corrosion Cracking. Mechanisms of Environment
     Sensitive Cracking of Materials. (London, The Metals So-
     ciety, 1977) 471.

8.   Kermani, M. and Scully, J.C. Fractographic Aspects of the
     Stress Corrosion Cracking of α-Brass in 15N Ammonia Solu-
     tions. Corrosion Science 19 (1979) 489-506.

9.   Mukai, Y., Watanabe, M. and Murata, M. Fractographic Ob-
     servations of Stress Corrosion Cracking of AISI 304 Stain-
     less Steel in Boiling 42 Percent Magnesium-Chloride Solu-
     tion. Fractography in Failure Analysis. ASTM-STP 645
     (Philadelphia, ASTM, 1978) pp 164-175.

10. Dickson, J.I., Groulx, D. and Tromans, D. Fractography of
     SCC of 310 Stainless Steel: Influence of K. To be pu-
     blished.

11. Scamans, G.M. and Swann, P.R., High Voltage Electron Metal-
     lography of Stress Corrosion Cracking of Austenitic Stain-
     less Steels. Corrosion Science 18 (1978) 983-995.

12. Dickson, J.I., Geckinli, E. and Uribe-Perez, I. Fractogra-
     phic Aspects of Cyclic Cleavage. Materials Science and
     Engineering 60 1983) 231-240.

13. Dickson, J.I., Lincourt, C., Haenny, L. and Baïlon, J.P.
     The Fatigue Crack Propagation of a Mild Steel Below the
     Ductile-Brittle Transition Temperature, presented at CFC9
     Montreal 1983, to be published.

14. Hanninen, H. and Hakkarainen, T. Fractographic Characte-
     ristics of a Hydrogen-Charged AISI 316 Type Austenitic
     Stainless Steel. Metallurgical Transactions 10A (1979)
     1196-1199.

15. Nakasato, F. and Bernstein, I.M. Crystallographic and
     Fractographic Studies of Hydrogen-Induced Cracking in
     Purified Iron and Iron-Silicon Alloys. Metallurgical
     Transactions 9 (1978) 1317-1326.

16. Bégin, R., Vandenbroucke, J. and Dickson, J.I., unpublished
     research, Ecole Polytechnique de Montréal, 1984.

17. Aitchison, I. and Cox, B. Interpretation of Fractographs
     of SCC in Hexagonal Metals. Corrosion 28 (1972) 83-87.

18. Dickson, J.I. and Tromans, D., unpublished research.

19. Martin, P., Dickson, J.I. and Baïlon, J.P., unpublished
     research, Ecole Polytechnique de Montréal, 1984.

20. Dickson, J.I., Baïlon, J.P. and Masounave, J. A Review on the Threshold Stress Intensity Range for Fatigue Crack Propagation. Canadian Metallallurgical Quarterly 20 (1981) 317-329.

21. Baïlon, J.P., Chappuis, P., Masounave, J. and Dickson, J.I. Fractographic Aspects of the Threshold in Several Alloys. Fatigue Thresholds: Fundamentals and Engineering Applications (London, Engineering Materials Advisory Services 1982) 277-291.

22. Baïlon, J.P., El Boujdaini, M. and Dickson, J.I. Environmental Effect on ΔK_{th} in 70-30 α-Brass and 2024-T3. TMS-AIME Symposium: Fatigue Crack Growth Threshold Concepts, in press.

23. Ait Bassidi, M., Dickson, J.I., Baïlon, J.P. and Masounave, J. The ΔK_{th} Behaviour of Three Stainless Steels in Different Environments, ICF6 Proceedings, in press.

24. Lindigkeit, J., Terlinde, G., Gysler, A. and Lütjering, G. The Effect of Grain Size on the Fatigue Crack Propagation Behaviour of Age-Hardened Alloys in Inert and Corrosive Environment. Acta Metallurgica 27 (1979) 1717-1726.

25. Beevers, C.J. Fatigue Crack Growth Characteristics at Low Stress Intensities of Metals and Alloys. Metal Science 11 (1977) 362-367.

26. Dickson, J.I., Handfield, L. and Baïlon, J.P., unpublished research, Ecole Polytechnique de Montréal, 1984.

27. Marchand, N. Fatigue oligocyclique et fatigue-propagation du cuivre et laiton-alpha. M. A. Sc. Thesis. Ecole Polytechnique de Montréal, 1983.

28. Boisson, P., Petit, J. and Gasc, C. Étude comparative de la vitesse de propagation de fissures de fatigue à l'air et sous vide dans les alliages légers à haute résistance. Mémoire Scientifique de la Revue de Métallurgie 74 (1977) 427-437.

29. Stubbington, C.A. Some Observations on Air and Corrosion-Fatigue of an Aluminium - 7.5% Zinc - 2.5% Magnesium Alloy. Metallurgia 68 (1963) 109-121.

30. Ait Bassidi, M., Masounave, J., Baïlon, J.P. and Dickson, J.I. Fractographic Study of Corrosion-Fatigue Crack Propagation in a Duplex Stainless Steel. Defects, Fracture and Fatigue (The Hague, Martinus Nijhoff 1983, 195-208).

31. Ait Bassidi, M., Masounave, J., Dickson, J.I. and Baïlon, J.P. Fatigue-corrosion de l'acier austéno-ferritique KCR 171 dans l'eau blanche. Canadian Metallurgical Quarterly, in press.

32. Ait Bassidi, M., Masounave, J. and Dickson, J.I. Corrosion-Fatigue Behaviour of Three Stainless Steels, presented at CFC9 Montreal, 1983, to be published.

33. Speidel, M.O. Influence of Environment on Fracture. Advances in Fracture Research (Oxford, Pergamon Press, 1982) 2685-2704.

34. Simmons, G.W., Pao, P.S. and Wei, R.P. Fracture Mechanics and Surface Chemistry Studies of Subcritical Crack Growth in AISI 4340 Steel. Metallurgical Transactions 9A (1978) 1147-1158.
35. Dickson, J.I., Handfield, L., Baïlon, J.P. and Sahoo, M. The Fatigue and Corrosion-Fatigue Crack Growth Behaviour of a Mn-Ni-Al Bronze. Proceedings of Fatigue 84, in press.
36. Kim, Y.H. and Morris, Jr., J.W. The Nature of Quasicleavage Fracture in Tempered 5.5 Ni Steel After Hydrogen Charging, Metallurgical Transactions 14A (1983) 1883-1888.
37. Caskey, Jr., C.A. Hydrogen-Induced Brittle Fracture of Type 304L Austenitic Stainless Steel. Fractography and Materials Science, ASTM STP 773 (Philadelphia, ASTM 1981) 86-97.
38. Gangloff, R.P. and Wei, R.P. Fractographic Analysis of Gaseous Hydrogen Induced Cracking in 18 Ni Maraging Steel. Fractography in Failure Analysis ASTM 645. (Philadelphia. ASTM, 1978) 87-106.
39. Bernstein, I.M., Garber, R. and Pressouyre, G.M. Effect of Dissolved Hydrogen on Mechanical Behaviour of Metals. Effect of Hydrogen on Behaviour of Materials (New York, TMS-AIME, 1976) 37-57.
40. Delebecq, J., Poudou, P., Ayel, J. and Pluvinage, G. Fatigue in a Lubricant Environment, presented at CFC9, Montreal 1983, to be published.

RESIDUAL STRENGTH OF A PRESSURIZED HEAT TRANSPORT PIPE CONTAINING
A DEFECT

Fernand Ellyin

Department of Mechanical Engineering
The University of Alberta
Edmonton, Alberta   T6G 2G8

ABSTRACT

The residual strength of a pipe (or a cylindrical vessel) containing a defect which is propagating in time, is predicted. The
pipe material and loading conditions are those existing in a heat
transport piping system of electric generating plants. Analytical
and experimental results are combined to predict the variation of
the residual strength in terms of either defect size or number of
applied cycles (or time). The residual strength decreases at a constant rate in the early life, and as the defect size gets larger,
the residual strength decreases rapidly.

INTRODUCTION

All structural components contain defects or cracks of varying
degrees even before their installation. They may vary from microscopic scale to those detectable by the present day techniques.
When these structural components are subjected to stresses or
strains, the absorbed energy may force the cracks to propagate.

In the first stage, the cracks are normally of a microscopic
scale so that their existence does not alter the macro-mechanical
characteristics of the material. However, as the component is subjected to further load applications, the cracks propagate, and the
ensuing damage may cause a reduction of the component strength. The
cracks or defects in this stage of the component life are generally
of a macroscopic scale. Subsequent applications of load would lead
to the crack growth; eventually reaching a critical crack length
which causes the failure of the component.

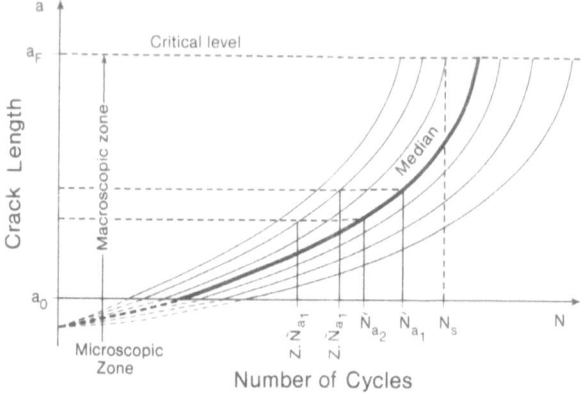

Fig. 1. Crack growth history of similar components.

The process of failure described above varies in time for a population of similar components. Figure 1 shows a schematic diagram of the crack propagation in a population at various stages. The philosophy of the fail-safe concept could be described by this figure. It relies on leaving the component or structure in service until the cracks (or defects) are detected by a planned inspection before they reach the dangerous level. A serious problem with this method is the assurance of sufficient residual strength to provide safety until the defects are detected. Several models have been proposed for predicting the safe life of a component in aviation as summarized by Payne (1).

One of the major problems to deal with is variability of the component strength with time. If the risk of failure under a pre-scribed load spectrum is desired, one would require a description of the variability in residual strength, see Ellyin (2). In this paper, an attempt is made to calculate the residual strength of a pressurized pipe (or vessel) containing a crack-like defect. In particular, we wish to determine the variation of strength with time or number of load applications. Such a relationship could then serve as a median distribution for a probabilistic study, Ref. (3).

RESIDUAL STRENGTH MODEL

Most theoretical and experimental works in fracture and crack propagation have been concerned with a flat plate configuration. The reason for this constraint is that both theoretical and experimental extensions to other configurations have proven difficult to solve and/or complex for practical applications. In a pressure vessel or piping system, the driving force for an axial crack propagation has at least two components. One is the hoop stress, and the other results from of the radial pressure which causes protrusion of the wall adjacent to the defect. The latter displacement mode does not occur in a flat plate.

Earlier tests carried out by NACA, Ref. (4), indicated the strong influence of the curvature and crack length. These tests also clearly demonstrated that the strength of a pressurized vessel is substantially less than that of a corresponding flat plate in tension. Consequently, Peter and Kuhn (4) proposed an empirical relationship which contained a curvature correction term. This relation has since been further developed and given theoretical backing by several researchers. The general concept is that the behavior of a cylindrical vessel can be predicted from a flat plate of the same material, thickness and length, and containing the same crack (Fig. 2) provided that the applied stress σ is taken to be:

$$\sigma = M \sigma_H \tag{1}$$

The multiplier M would be a function of crack length 2a, the mean radius R_m, thickness t, and material properties. For a crack propagation, the required hoop stress can be expressed in terms of the nominal stress in the flat plate σ^* (Fig. 2), i.e.,

$$\sigma^*_H = \sigma^* M^{-1} \tag{2}$$

The above relation is attractive since it permits the utilization of a large body of fracture toughness data existing for flat plates. Relation (2) is further extended to include the effects of plastic flow at the crack region. Another advantage of this type of formulation is that it lends itself to making predictions.

Different expressions for the multiplier M are proposed, where the major ones are summarized in Table 1. Folias' expression (5) is based on analytical derivation of stress fields and linear elastic fracture mechanics (LEFM).

Anderson and Sullivan, Ref. (6), employed the linear elastic fracture mechanics concept by substituting for σ^* (which depends on crack length), the material fracture toughness K_c (which is independent of crack length). The function $\phi = \phi (\sigma^*/\bar{\sigma})$ is a plasticity correction to the LEFM, and β is a function of material and pipe geometry. It is to be noted that K_c varies with the plate thickness and that the K_c value to be used with the expressions given in Table 1 should be compatable with the wall thickness of the pipe or vessel. A formula was obtained by Crichlow and Wells (7) from fitting experimental data to the available forms of analytical expressions to arrive at an empirical relation listed in Table 1.

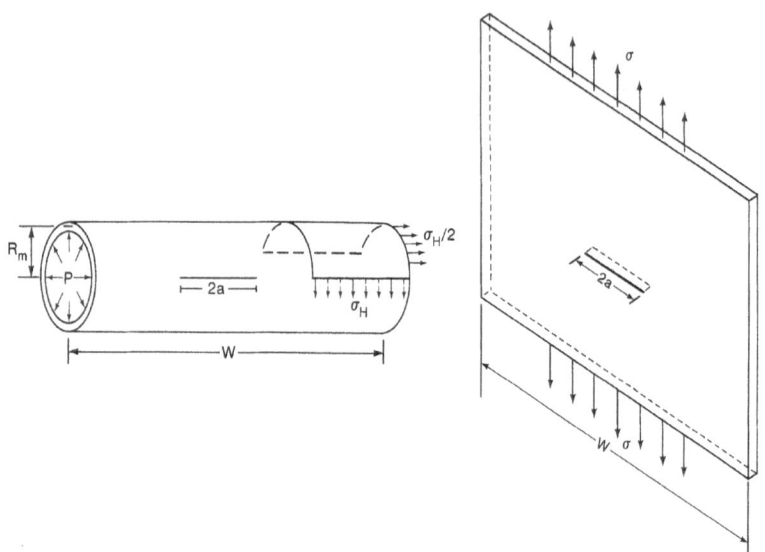

Fig. 2. A pressurized pipe (or vessel) containing an axial flaw and corresponding flat plate subjected to a stress $\sigma = M\sigma_H$

TABLE 1 CRITERIA FOR CRACK PROPAGATION IN PRESSURIZED PIPES OR CYLINDRICAL VESSELS

FAILURE CRITERION	M	ϕ	REF.
$\sigma_H^* = \sigma^* M^{-1}$	$(1 + 9.2\ a/R_m)$	-----	(4)
$\sigma_H^* = \sigma^* M^{-1}$	$[1 + 1.61(a^2/R_m t)]^{1/2}$	-----	(5)
$\sigma_H^* = \dfrac{K_c\ M^{-1}}{(\pi a \phi)^{1/2}}$	$(1 + \beta a/R_m)$	$\left[1 + \dfrac{(M\sigma_H^*)^2}{2\sigma_y^2}\right]$	(6)
$\sigma_H^* = \sigma^* M^{-1}$	$[1 + 0.81\ (\dfrac{a}{(R_m t)^{1/2}})^{3/4}]$	-----	(7)
$\sigma_H^* = \dfrac{K_c\ M^{-1}}{(\pi a \phi)^{1/2}}$	$[1 + 1.61\ a^2/R_m t]^{1/2}$	$\sec\dfrac{\pi\sigma_H^*}{\sigma_y + \sigma_u}$	(8, 9)

Duffey, McClure, Eiber, and Maxey's (8, 9) proposed relation-
ship is similar to that of Ref. (6) except that Folias' equation for
M is used, and ϕ is derived from the Dugdale's crack model (10).
The expression derived by Duffy et al. is shown to be in good agree-
ment with various experimental data. Certain limitations of the
Duffey et al. criterion has been discussed in Ref. (11, 12).

Hahn, Sarrate, and Rosenfield (11, 12) suggested improvement
to Duffey et al. criterion by incorporating a more rigorous plasticity correction based on the crack-tip displacement. When this correction is employed, it becomes apparent that the fracture toughness
plays a minor role in the extension of short cracks in tough
materials. Hahn et al. (11, 12), in essence, proposed three closely
related criteria for the extension of axial cracks in pressurized
cylindrical vessels or pipes: (i) a fracture-toughness criterion
mainly for low- and medium-tough materials, (ii) a flow-stress
criterion for short cracks in tough materials, and (iii) a modification of (i) for very thin vessels. These three criteria are described in Table 2. The predictions of these criteria have been
compared with a large body of experimental results, and have been
shown to be in fairly good agreement.

FATIGUE CRACK PROPAGATION

In the previous section, it was shown that the residual
strength of a pipe containing a crack and subjected to monotonic
loading could be estimated from the expressions given in Tables 1
and 2. In the case of cyclic loading, the fatigue crack growth

TABLE 2 CRITERIA FOR CRACK PROPAGATION, REF. (11, 12)

$5 < R_m/t < 50 \rightarrow M = (1 + 1.61\, a^2/R_{iii}t)^{1/2}$		
$R_m/t > 50 \rightarrow M = \{1 + 1.61(a/R_m)^2[50\ \tanh(R_m/50t)]\}^{1/2}$		
$\sigma_H^* = K_c\,(\pi a M^2)^{-1/2}$	$\sigma_H^* = \dfrac{K_c}{(\pi a \phi_3)^{1/2}}\,M^{-1}$	$\sigma_H^* = \bar\sigma M^{-1}$
	$\phi_3 = \left(\dfrac{\pi\sigma_H^* M}{2\bar\sigma}\right)^{-2} \ln\left(\sec\dfrac{\pi\sigma_H^* M}{2\bar\sigma}\right)^2$	
long cracks	$\sigma_y < \bar\sigma < \sigma_u$	short cracks
≈ 1.2	≈ 7	$\dfrac{1}{a}\,(K_c/\sigma_y)^2$

rate, da/dN, may be described by a sigmoidal curve in log da/dN versus log ΔK coordinate axes. In the intermediate range of stress intensity factor, ΔK, it can be shown that log da/dN is almost linearly related to log ΔK, i.e.

$$\frac{da}{dN} = C(\Delta K)^m \qquad (3)$$

where C and m are related to material properties. This relationship was first proposed by Paris and Erdogan (13). Recently, it has been shown by Kujawski and Ellyin (14), that relation (3) is a particular case of a general crack propagation law, and that parameters C and m are interrelated. To obtain residual strength of a pressurized pipe (or cylindrical vessel), under cyclic loading Eq. (2) may be written as:

$$\Delta\sigma_H = \Delta\sigma \ M^{-1} \qquad (4)$$

where $\Delta\sigma$ is now the cyclic stress amplitude. For a plate with a central crack, Fig. 2, the stress amplitude $\Delta\sigma$ is related to the range of stress intensity factor, ΔK by (15)

$$(\Delta K)^2 = (\Delta\sigma)^2 W \tan (\pi a/W) \qquad (5)$$

Combining equations (3), (4) and (5), one can determine the residual strength of the pipe in terms of crack length, provided the material properties and geometric parameters are known. It is shown that m and C in Eq. (3) are functions of fatigue properties, cyclic stress-strain exponent, modulus of elasticity, and a material length parameter (14). They also vary with the mean stress and temperature (16).

APPLICATION

A piping system in a power plant or process industry will be considered. The pipe carries pressurized hot fluid at about 316°C (600°F) and is constructed from a low alloy carbon steel. To carry out the analysis, we would require the crack propagation parameters of Eq. (3). The crack growth rate in carbon steels is very sensitive to the temperature, and therefore, the room temperature data cannot be used. No analytical methods are available to correlate room temperature data to a higher one. For any realistic prediction of the heat transport piping system, the material properties have to be obtained through experiments at the operating temperature and loading range. In recognition for this need, a number of experiments on fatigue crack growth of piping steels were conducted by the General Electric Company for the former U.S. Atomic Energy Commission (AEC) Ref. (16). These experiments reflect the material condition and temperature at the operating range of the heat transport piping systems. The experimental results of Brothers (16) on A-212B piping steel at 316°C (600°F) with a gross stress of 198.6 MPa (28 800 psi) applied at a frequency of 0.6 cpm were fitted to the power law relation (3) by a least squares method. The result was $C = 6.6181467 \times 10^{-10}$ and exponent $m = 2.9535877$. Substituting from (5) into (3) we now obtain the relationship between crack propagation rate and the applied stress range for the experiments of Ref. (16),

$$\frac{da}{dN} = 6.6181467 \times 10^{-10} \{\Delta\sigma[W \tan (\frac{\pi a}{W})]^{1/2}\}^{2.9535877} \qquad (6)$$

To obtain the number of cycles for a defect to propagate from a given length to another one, it will suffice to integrate Eq. (6). Of course, the integration has to be carried out numerically, and for this, the stress range $\Delta\sigma$ and width W have to be specified.

NUMERICAL RESULTS AND COMPARISON

Due to the nature of governing equations in Tables 1 and 2, different methods of solution were required to obtain residual strength predicted by each theory. Computer programs were prepared to solve the functional relationship of Hahn et al. (11, 12) and Duffey et al. (8, 9). Two limiting cases were studied for a pipe with 508 mm (20 in.) diameter: a thin walled R/t = 100 and a thick one R/t = 10. These two limits define boundary of theoretical transition from a plane stress condition to the plane strain one.

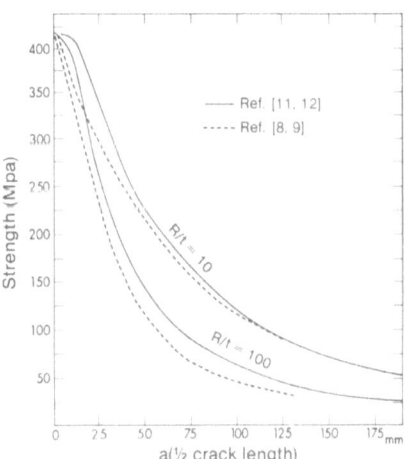

Fig. 3 Residual strength versus crack length for thin- and thick-walled pressurized pipes

The other material parameters were K_c = 130 MPa $\sqrt{m}$ (118 ksi $\sqrt{in}$) and $\bar{\sigma} = (\sigma_y + \sigma_u)/2$ = 419.5 MPa (60.85 ksi). The results are presented in Fig. 3. One notes the pronounced effect of thickness t on the residual strength. The other observation is that for a given stress level, the relative critical crack length is smaller for a thicker pipe. The two theories indicate the same tendency with respect to the crack length. Either theory could therefore be used for the subsequent study. However, it is to be noted that Duffey et al.'s functional form is easier to calculate than that of Hahn et al.

To obtain variation of the residual strength with time, Eq. (6) has to be integrated and substituted into Eq. (4). To generalize the results, we define the following dimensionless parameters. The relative life is defined by:

$$V = \frac{N}{N_i} \tag{7}$$

where N is number of applied cycles and N_i is the number of cycles to initiate a fatigue crack. Similarly, the relative crack length is expressed by

$$h = \frac{2a}{2a_f} \tag{8}$$

where a_f is the critical length. The relative crack length h in terms of the relative life for the pipe material was calculated from Eq. (6) to (8) with $\Delta\sigma$ = 196.5 MPa(28 500 psi), W = 3 048 mm(120 in.). It was noted that for the relative lives V > 6, the crack growth rate increased rapidly. No single equation could accurately describe the obtained numerical data. They were thus divided into three parts and polynominal up to 16th degree were fitted by the least squares technique.

The relative residual strength is defined as

$$\phi(h) = \frac{\mu_R(h)}{\mu_0} \tag{9}$$

where $\mu_R(h)$ is the residual strength of the component with the relative crack length h, and μ_0 is the strength of the uncracked pipe (h = 0). It is convenient to express the residual strength in terms of the hoop stress σ_H, thus making the results independent of the geometry of the pipe. In this case, μ_0 becomes the ultimate strength of the material in tension. Substituting in (9) for $\mu_R(h)$ from Eq. (4 to 9), we obtain the relationship between the relative residual strength $\phi(h)$ and the relative life V, as shown in Fig. 4. It is seen that residual strength decreases as the component life in service increases. After a certain life, V > 5, the rate of

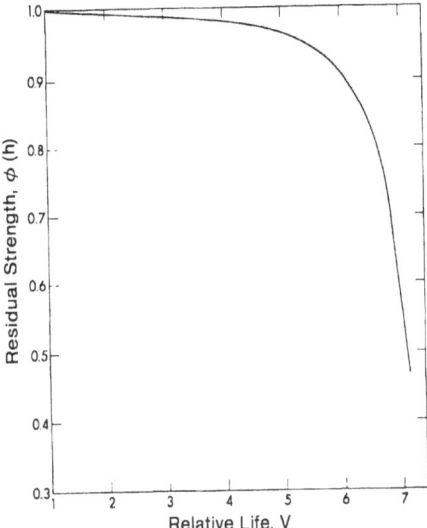

Fig. 4   Variation of the relative residual strength with relative
         life of a pressurized pipe.

decrease in the residual strength is accelerated, eventually leading
to failure by fast fracture.

CONCLUSIONS

The residual strength of a pipe carrying pressurized hot fluid
at 316°C (600°F) has been determined in terms of defect length, or
the number of applied cycles (time). Inversely, for a given crack
length, the critical stress at which the crack propagation would
commence has been determined. The assurance of sufficient residual
strength is central to providing safety until defects are detected.
It is noted that variations in the stress range and/or thickness to
diameter ratio produce curves of a similar shape for either crack
length versus number of cycles, or the residual strength in terms of
applied cycles.

REFERENCES

1.   Payne, A.O. A Reliability Approach to the Fatigue of Structures
     (Probabilistic Aspects of Fatigue, ASTM, STP 511), (American
     Society for Testing and Materials, 1972) pp. 106-155.
2.   Ellyin, F. A Strategy for Periodic Inspection Based on Defect
     Growth - Part I - Probabilistic Formulation, (Theoretical and
     Applied Fracture Mechanics) (submitted to).

3. Ellyin, F. A Strategy for Periodic Inspection Based on Defect Growth - Part II - Application to a Piping System (Theoretical and Applied Fracture Mechanics) (submitted to).

4. Peters, R.W. and P. Kuhn. Bursting Strength of Unstiffened Pressure Cylinders With Slits, (National Advisory Committee for Aeronautics, NACA), Technical Note 3993, Washington, (April, 1957), 20 pages.

5. Folias, E.S. An Axial Crack in a Pressurized Cylindrical Shell, (International Journal of Fracture Mechanics), 1 (1965), pp. 104-114.

6. Anderson, R.B. and T.L. Sullivan. Fracture Mechanics of Through-Cracked Cylindrical Pressure Vessels, (National Aeronautics and Space Administration NASA), Report No. TND-3252, Washington, (1966).

7. Crichlow, W.J. and R.H. Wells. Crack Propagation and Residual Static Strength of Fatigue-Cracked Titanium and Steel Cylinders (Fatigue Crack Propagation, ASTM, STP-415), (American Society for Testing and Materials, 1967), pp. 25-70.

8. Duffey, A.R., McClure, G.M. Eiber, R.J. and W.A. Maxey. Fracture Design Practices for Pressure Piping, Fracture, in H. Liebowitz ed., (Academic Press, 1969) Vol. 5, pp. 159-232.

9. Duffey, A.R. Eiber, R.J. and W.A. Maxey. Recent Work on Flaw Behaviour in Pressure Vessels, (Proceedings of the Symposium on Fracture Toughness Concepts for Weldable Structural Steel), Risely, U.K., (April, 1969), pp. M1-M34.

10. Dugdale, D.S. Yielding of Steel Sheets Containing Slits, (Journal of the Mechanics and Physics of Solids), 8 (1960) pp. 100-104.

11. Hahn, G.T., Sarrate, M. and A.R. Rosenfield. Criteria for Crack Extension in Cylindrical Pressure Vessels, (International Journal of Fracture Mechanics), 5 (1969) pp. 187-209.

12. Hahn, G.T. and M. Sarrate. Failure Criterria for Through-Cracked Vessels, (Proceedings of the Symposium on Fracture Toughness Cocnepts for Weldable Structural Steel), Risley, U.K., (April, 1969), pp. P1-P15.

13. Paris, P.C. and F. Erdogan. A Critical Analysis of Crack Propagation. Journal of Basic Engineering, Trans. ASME, 85 (1963), pp. 528-534.

14. Kujawski, D. and F. Ellyin. A Fatigue Crack Propagation Model, (Department of Mechanical Engineering, University of Alberta, Edmonton, Alberta), (1984).

15. Rooke, D.P. and D.J. Cortwright. Compendium of STress Intensity Factors, (Her Majesty's Stationery Office, London, U.K. 1976).

16. Brothers, A.J. Fatigue Crack Growth in Nuclear Reactor Piping Steels, (AEC Research and Development Report, GEAP-5607), General Electric Co., California, (March 1968), 52 pages.

MICROSTRUCTURAL EFFECTS ON CREEP AND FATIGUE

M. E. Fine and J. R. Weertman

Department of Materials Science and Engineering and Materials
Research Center, Northwestern University, Evanston, IL 60201 USA

ABSTRACT

Both creep and fatigue are highly complex phenomena involving
many factors. The roles of microstructure are even more complex
and often conflicting. This paper discusses the roles of microstructure in low temperature phenomena first, that is, fatigue at
room temperature and below, then elevated temperature creep, and
finally fatigue again but at elevated temperatures. At low temperatures fatigue crack initiation, microcrack propagation, the stress
intensity range threshold for macrocrack propagation including
closure stresses, and mid-range macrocrack propagation are discussed
in sequence. The discussion of creep and high temperature fatigue
emphasizes the influence of microstructure on damage accumulation
at grain boundaries and microstructural instability during service.

FATIGUE AT ROOM TEMPERATURE AND BELOW

The fatigue failure process may be divided into several stages:
1) initiation of one or more microcracks, 2) propagation or coalescence of microcracks to form one or more macrocracks and 3) propagation of one or more macrocracks to final failure. A very early
concept in the science and art of fatigue of metals is that the
microstructure became "tired" from cyclic loading and the metal
failed by a cyclic "aging" process (1). The accumulation of microstructural damage to cause initiation of a fatigue crack is highly
localized and the "aging" must be highly localized. Theories of
fatigue failure based on general strain or energy accumulation
throughout the specimen cannot be correct.

Fatigue cracks usually initiate near or at singularities which
lie on or just below the surface. Such singularities may be sharp

changes in cross-section, pits, inclusions, embrittled grain boundaries, etc. However, even when the surfaces of metals are highly
polished, the metal is flaw free and no stress concentrators are
present, a fatigue crack may still form. Localized regions of
plastic deformation develop under continued cycling until they become sufficiently severe to initiate one or more fatigue cracks.
Such cracks, when initiated, are very small. The initially observed fatigue cracks have become smaller and smaller as the resolving
power of the microscopes used to study them have become smaller.
With TEM of surface replicas taken with a load applied to open the
crack, cracks as small as 0.1 μm have been observed along slip
bands (2). Observation of such small fatigue cracks is quite general (3). Figure 1 (4) is an example showing two tiny cracks adjacent to a slip band in OFHC copper cycled 100 times at a plastic
strain amplitude of 5×10^{-4}.

How do such small fatigue cracks form? One suggestion is that
they form by accumulation of vacancies generated by the to and fro
dislocation motion (5). These might be imagined to diffuse along
dislocation pipes until they reach the surface forming a pit where
the dislocation emerges. Since tiny fatigue cracks also form at
4.2° K (6), such a mechanism or any thermally activated process
cannot have general applicability.

Some success has been achieved with a model (4) based on vacancy type dislocation dipole accumulation in the slip band. A
crack is proposed to initiate when the accumulated displacement in
the dipole pile-up equals the critical strain for fracture of a
perfect metal. The model predicts that the crack should initiate
near the slip band matrix boundary, as observed with OFHC copper.
The resulting equation for the cycles to initiation is

$$N_i \cong \frac{0.4}{\pi(1-\nu)} \ \frac{\mu^2 k}{(\Delta\tau-2k)(\Delta\tau)^2} \ \frac{b}{a} \ \frac{(1-R)^2}{f^*} \ m^3 \tag{1}$$

where ν = Poisson's ratio, k = resistance stress to dislocation
motion, $\Delta\tau$ = stress range, b = Burger's vector, a = half slip band
length, R = stress ratio, m = orientation factor, e.g., Taylor
factor, and f^* = damage accumulation efficiency factor, i.e.,
average ratio of dislocations stored per cycle to dislocations
stored during first half cycle.

In this equation k, a, and f^* depend on microstructure. As
the yield strength or k, which is a function of microstructure,
increases, an increase in N_i is predicted as generally observed.
The grain size influences N_i not only through k but also through a
since a is expected to increase with grain size. Lin et al. (4)
estimated f^* from the stored energy and also from the back stress
obtained from hysteresis loops. If the dislocation motion is

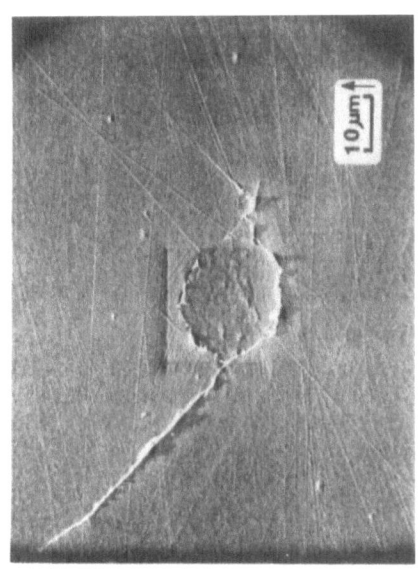

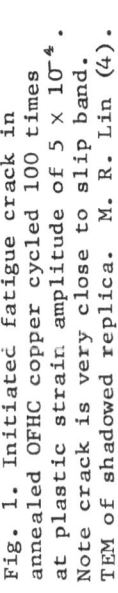

Fig. 2. Initiation of fatigue cracks near inclusions in Fe-3% Si. Inclusions became debonded during cycling. The sample was replicated after 3000 cycles at a strain amplitude of 5×10^{-4} and the replica was examined in the SEM. A large and small inclusion are shown side by side. A long fatigue crack is observed approximately 45° to the stress axis indicated by the arrow near the micron marker.

Fig. 1. Initiated fatigue crack in annealed OFHC copper cycled 100 times at plastic strain amplitude of 5×10^{-4}. Note crack is very close to slip band. TEM of shadowed replica. M. R. Lin (4).

reversible then f* is small. Thus f* is also a function of micro-structure. In general, soft metals will have large values of f* and low N_i. The net prediction is that N_i will increase rapidly with increase in yield stress.

When inclusions are present, they hasten fatigue crack initi-ation particularly in the long-lifetime regime. Such inclusions act as local stress risers which reach larger maximum values as the yield stress increases. Thus fatigue cracks are expected to in-creasingly form at or adjacent to inclusions as the yield stress increases. Even if the inclusion is cracked or debonded from the matrix, the crack must extend into, that is, initiate in the matrix. Fatigue cracks adjacent to inclusions often form on slip bands eman-ating from the inclusion (7). Figure 2 shows a fatigue crack along a slip band emanating from an inclusion in 3% Si-Fe (4). Here the inclusion is debonded from the matrix. In other cases fatigue cracks emanating from inclusions are normal to the stress direction. Such fatigue cracks are suggested to form by interaction of dislo-cations on two slip bands intersecting at the inclusion matrix interface similar to the Stroh mechanism of crack formation. The probability that an inclusion initiates a fatigue crack decreases with the size of the inclusion (7). It is suggested that inclusions $\frac{1}{2}$ μm in size or less play little role in enhancing fatigue crack initiation at room temperature unless the inclusions are segregated.

Summarizing the effects of microstructure on fatigue crack in-itiation, the important microstructural features are slip bands and inclusions. Thus increase in yield strength increases the stress to cause slip bands but the material is less likely to relieve stress concentrations at singularities, such as inclusions, by plas-tic flow. However, we still don't know the microstructural processes by which fatigue cracks form at room temperature and below.

Once microcracks form, for failure to occur they must grow and/ or coalesce to form macrocracks. The process by which microcracks develop into macrocracks is discontinuous. There is much evidence that microcrack growth is impeded by grain boundaries (7,8). Thus small grain size may be beneficial for slowing microcrack growth as well as for increasing resistance to fatigue crack initiation. Since microcrack growth is discontinuous, there is no way that fracture mechanics alone can be used as the predictor of fatigue crack growth rate in the microcrack growth region for a given material.

In the short crack regime, the crack growth rate often has been observed to decrease as the crack extends under application of cycles at constant load range, $\Delta\sigma$ (9,10). The usual procedure is to initiate the crack at a notch in a panel specimen at constant $\Delta\sigma$ and then begin monitoring of the crack length when it appears on the sides of the panel. Some of the initial rapid rate is due to the crack being initially thumbnail in shape, but even if this is corrected for, a minimum in the da/dN vs. ΔK curve is observed, as shown in Fig. 3a (11). However, crack closure seems to be mainly responsible for this effect for if da/dN is plotted vs. ΔK_{eff}, i.e., $K_{max} - K_{closure}$, the effect essentially disappears, as shown in Fig. 3b (11). Crack closure and its relation to microstructure will be discussed subsequently.

Once a macrocrack has formed, i.e., a crack considerably larger in both dimensions than the grain size, then the log da/dN vs. log ΔK curve is divided into the customary three stages: I, near threshold; II, the Paris relation region; and III, near final failure. Stage III will not be discussed except to say that da/dN varies with microstructure in a similar way to K_c.

In the near threshold region the fatigue crack propagation rate is much affected by microstructure such as grain size (12); however, most of the grain size effect in vacuum melted iron disappears if da/dN is plotted versus ΔK_{eff} rather than ΔK (13). The latter study was done in dry argon atmosphere so corrosion fatigue is not playing a role. Following Suresh and Ritchie (14), such crack closure is attributed to fracture surface roughness coupled with shear displacements. Their equation is

$$K_{cl}/K_{max} = \sqrt{(2\gamma x)/(1 + 2\gamma x)} \qquad (2)$$

where γ is the ratio of the average surface roughness to the wavelength of the surface roughness ($\bar{H}/w$) and x is the ratio of mode II to mode I displacements. Near threshold, mode II fracture is observed and the proportion increases as the threshold stress intensity range, ΔK_{th}, is approached (15). This was quantified in a recent study and x was assumed to be equal to the areal fraction of mode II fracture to mode I fracture (16). Surface roughness was also measured and w was taken to be equal to the grain size. Excellent agreement (16) with eq.(2) was obtained for vacuum melted iron and Al-3% Mg alloy with low R($\sigma_{min}/\sigma_{max}$) ratios. Thus K_{cl}/K_{max} increases with the product γx which in turn is a function of microstructure. For example, the slip distances increase with grain size giving greater surface roughness.

The question of whether ΔK_{th}^{eff} is a function of microstructure or not remains to be discussed. The small increase with grain size in vacuum melted iron has already been mentioned (13). Dual phase microstructures in 1018 and HY80 steels where martensite is the continuous phase give higher ΔK_{th}^{eff} values than conventional heat treatments (17,18,19). Also, ΔK_{th}^{eff} in Al-3% Mg is larger at 77° K than 298° K (16). Any theory of ΔK_{th}^{eff} must account for these facts. It was previously suggested (20) that since fatigue crack

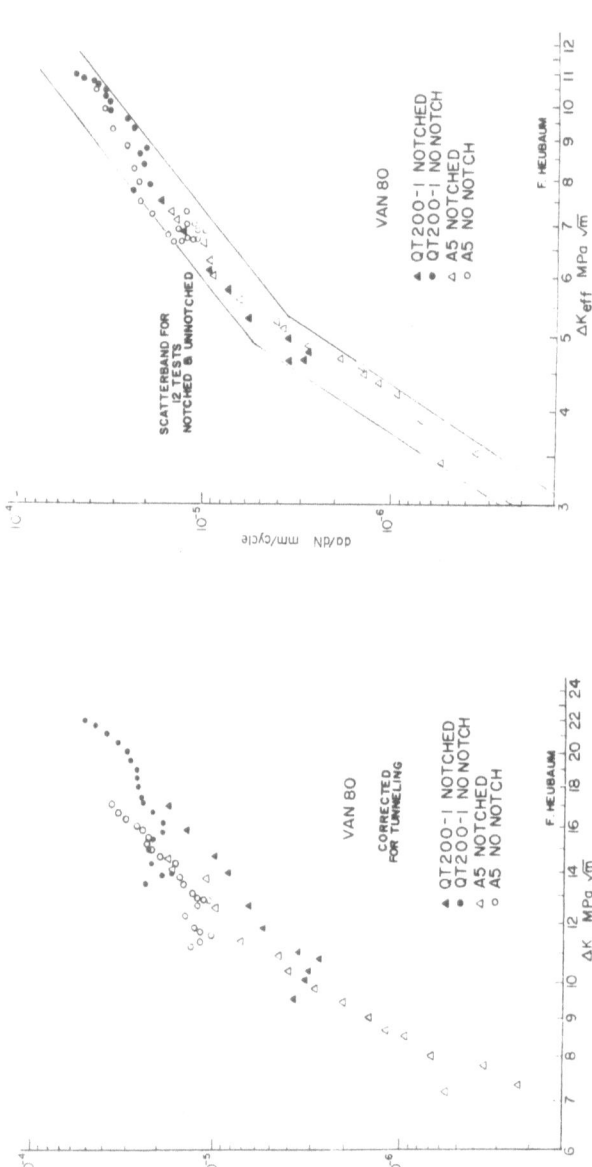

Fig. 3. Crack length vs (a) ΔK and (b) ΔK_{eff} for short cracks (0.2 to 1 mm long) in annealed and quenched and aged VAN 80 steel. Cracks were monitored on sides of notched and unnotched panel specimens. The notched specimen contained a 150 μm deep notch to assist crack initiation. In the no notch specimens, the notch was machined away after the crack had been initiated. Tests were conducted at constant nominal stress amplitude of approximately 240 MPa. ΔK is corrected for curvature of the crack front (tunneling) when it occurred.

propagation occurs by processes requiring dislocation motion, the stress to activate a dislocation source near the crack tip may determine ΔK_{th}. This stress is expected to increase with yield stress. The stress at the crack tip is relieved by plastic flow at the crack tip and thus such stresses are lower in soft materials. These two factors counteract each other. In dual phase steels the combination of hard martensite to give high yield stress and soft ferrite to reduce the stress concentration may yield an optimum microstructure giving resistance to fatigue crack propagation in the near threshold region.

In region II of the da/dN vs. ΔK curve, many people believe that changing microstructure has little effect; however, on thermodynamic grounds (21) an equation of the form

$$\frac{da}{dN} = \frac{A(\Delta K)^4}{\mu \sigma_y^{\prime 2} U} \tag{3}$$

may be derived. In eq.(3) μ is the shear modulus, $\sigma_y^{\prime}$ is an appropriate measure of the cyclic yield strength, U is the energy to form a unit area of crack, and A is a constant. Since U is a function of ΔK, other values than 4 for the exponent in the Paris relation are accounted for. The parameter U has been measured by several methods for various values of ΔK (21). For a set of 12 aluminum and iron base alloys, many with several different heat treatments, A was found equal to $2.9 \pm 0.8 \times 10^{-3}$ using International Units with 2.9 being the mean value and 0.8 being the standard deviation. The range of da/dN covered in the data set was 0.16 to 32×10^{-8} m/cycle, $\sigma_y^{\prime}$ varied from 42 to 690 MPa, and U from 0.17 to 53×10^5 J/m². The shear modulus of steel is three times that of Al alloys. Thus eq.(3) appears to be rather well confirmed.

The insensitivity of $|da/dN|_{\Delta K}$ to alloying and microstructure which is frequently observed is due to an inverse relationship between U and $\sigma_y^{\prime}$. The local plastic work is integrated throughout the plastic zone to obtain U and as $\sigma_y^{\prime}$ increases, the plastic zone size decreases tending to decrease U. The local plastic work density for the coordinates X and Y in the plastic zone is approximately

$$U_{XY} = \frac{|\sigma^{\ell} \epsilon^{\ell}|_{XY}}{|da/dN|} \tag{4}$$

where σ^{ℓ} is the maximum local stress and ϵ^{ℓ} is the local plastic strain range. Even though σ_{XY}^{ℓ} increases with $\sigma_y^{\prime}$ because the plastic zone can sustain higher stresses, U_{XY} for constant value of X and Y is usually smaller for higher strength alloys because ϵ_{XY}^{ℓ} is smaller. Liaw et al. (22) have given the following example. HY130 steel has a higher strength and lower U than hot rolled Nb-HSLA steel and yet the values of da/dN at ΔK of 20MN/m$^{\frac{3}{2}}$ are comparable. For $X = 100$ μm and $Y = 30$ μm, σ_{XY}^{ℓ} for HY130 is 540MN/m^2 and ϵ_{XY}^{ℓ} is 0.012. On the other hand, for the hot-rolled Nb-HSLA steel, at the same X and Y, σ_{XY}^{ℓ} is 270MN/m^2 and ϵ_{XY}^{ℓ} is 0.094.

In order to design alloys with improved resistance to fatigue crack propagation we need to know how to increase ϵ_{XY}^{ℓ}. The major contribution to U is the hysteretic plastic work associated with the back and forth dislocation motion. The hysteresis loop width versus stress range $\Delta\sigma$ in uncracked-unnotched specimens, a much easier set of measurements to obtain than those to determine ϵ_{XY}^{ℓ} and U, should indicate how σ_{XY}^{ℓ} and ϵ_{XY}^{ℓ} vary in the plastic zone ahead of a fatigue crack. Such a study versus microstructure may be helpful towards developing alloys which have low fatigue crack propagation rates.

DEFORMATION AT ELEVATED TEMPERATURES

Let us now examine what happens to metals subjected to deformation when the temperature T rises to a value such that diffusion of vacancies becomes important (T $\geq$ $\sim 0.4T_m$, where T_m is the melting temperature). In many cases grain boundaries, which in a sense can be regarded as a source of strength at low temperatures (e.g., Hall-Petch behavior and ductile-to-brittle transition temperature in steel), become the path of failure. Some 65 years ago Jeffries (23) advanced the concept of the equicohesive temperature T_{EQ}. It was suggested that at temperatures around T_{EQ} the strength of the matrix is approximately equal to that of the grain boundaries. Above T_{EQ} the boundaries are the weaker of the two components. While the picture of an equicohesive temperature is somewhat outmoded, it certainly is the case that at high temperatures intergranular failure becomes the dominant failure mode for many metals. In particular, materials under stress at high temperature are prone to grain boundary cavitation. Small voids nucleate on certain grain boundaries, grow under continued deformation, coalesce to form microcracks and eventually cause failure. The phenomenon of grain boundary cavitation occurs in a large number of metals and alloys, under both cyclic and time independent (creep) loading. While it has been observed that voids can nucleate in high purity metals containing no hard grain boundary particles (24), if such particles are present they are the favored void nucleation sites (25,26). For example, Fig. 4 shows a void which has nucleated during creep at a SiO_2 particle in an internally oxidized specimen of copper alloyed with a small amount of silicon. Raj and Ashby (27) have shown that the critical volume for void stability under a given stress acting across the grain boundary is greatly reduced if the void forms at the tip of a grain boundary particle. The actual stress assisting in the nucleation process may be enhanced by a substantial amount as the result of stress concentrations around a second phase grain boundary particle.

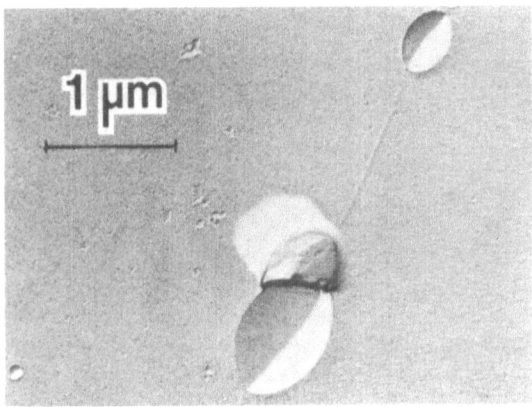

Fig. 4. A void which has nucleated and grown out from an SiO$_2$ boundary particle, in Cu-SiO$_2$ crept at 550°C and 12.4 MPa to about 5% of
its life. From (26).

The rate of void nucleation $\dot{N}$, and thus the rate of damage accumulation during high temperature creep or fatigue is markedly affected
by the microstructure of the material, especially the nature and
geometry of its grain boundary particles.

Creep Damage

Internally oxidized materials are well suited to an investigation
of the influence of microstructure on void nucleation rates (26).
By changing the alloy content and oxidizing conditions it is possible
to produce a wide variation in the average size and average spacing
of the oxide particles. The effect of interfacial energy on the
nucleation rate can be studied by changing the alloying element.
Even the influence of particle shape on $\dot{N}$ can be noted by internally
oxidizing a single crystal and then causing it to recrystallize before the high temperature mechanical testing is carried out. (The
oxides which form at the grain boundaries generally have a different
shape from those which nucleate and grow in the matrix.) Figure 5
shows the relationship between "cavitation level" (CL) and the average fraction of grain boundary area covered by particles, p/λ. The
dependence of CL on interparticle edge-to-edge spacing, λ', is
given in Fig. 6. Here CL is the percentage of grain boundary particles seen to be associated with a void, p is the average apparent
diameter of the grain boundary particles, and λ is their average
apparent center-to-center spacing. All of these quantities were
measured from shadowed two stage replicas taken from crept or
fatigued specimens of Cu-SiO$_2$. It can be seen from Figs. 5a and 6a

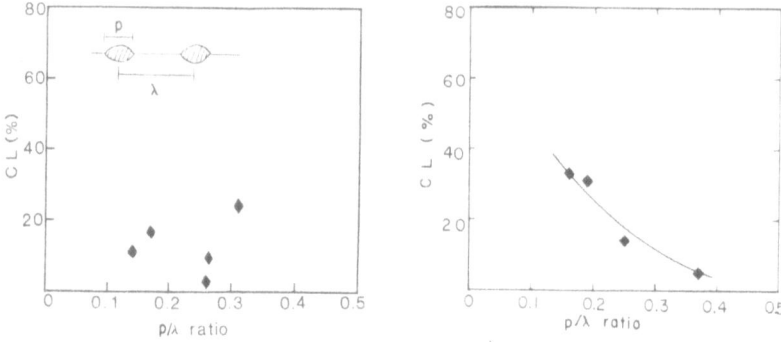

Fig. 5. Dependence of CL on the ratio p/λ where p is the average (apparent) particle size and λ is the average (apparent) center-to-center particle spacing: (a) Cu-SiO$_2$ crept at 550°C and 12.4 MPa to 0.20% strain; (b) Cu-SiO$_2$ fatigued in reverse bending at 405°C and 17 Hz, with a strain amplitude of 0.040% for 3×10^4 cycles. From (26).

that there is no obvious correlation between CL and p/λ in crept material whereas a direct relationship between CL and λ' is evident. These results seem to indicate that the stress concentration at the intersection of a hard particle and the grain boundary which appears when grain boundary sliding takes place does not play a dominant role in the nucleation of creep cavities. (Note that the stress acting on a grain boundary particle is concentrated by the factor

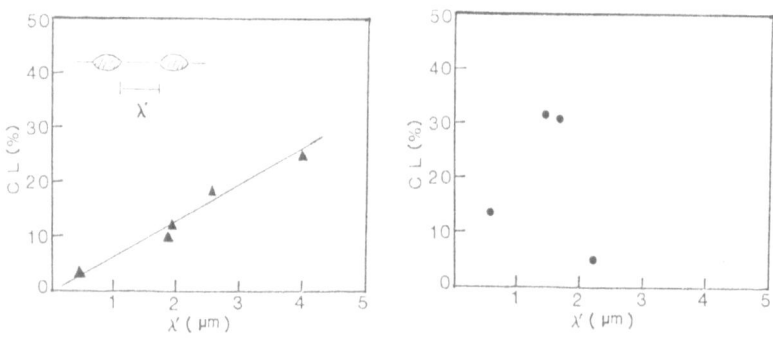

Fig. 6. Dependence of CL on the average edge-to-edge interparticle spacing λ': (a) Cu-SiO$_2$ crept at 550°C and 12.4 MPa to 0.20% strain; (b) Cu-SiO$_2$ fatigued in reverse bending at 405°C and 17 Hz, with a strain amplitude of 0.040% for 3×10^4 cycles. From (26).

p/λ if the shear stresses across the boundary are completely relaxed.) Such stress concentrations are important in theories of
void nucleation by a classical vacancy clustering mechanism (27,28).
On the other hand, the clear relationship between interparticle
spacing and CL lends support to a dislocation pile-up model of
creep cavity nucleation such as that proposed by Fleck et al. (25).

A comparison of cavitation in internally oxidized Cu-Si and
Cu-Ti alloys (26) shows that, as expected, void nucleation occurs
much more readily in the Cu-SiO$_2$ material. The interfacial energy
of the oxide particles in Cu-TiO$_2$ is approximately 0.4-0.7 J/m^2 (29),
which is about one-half of that of the Cu-SiO$_2$ particles (about 1.0-
1.2 J/m^2 (30)). It appears that the shape of the particle influences ease of nucleation in the manner predicted by theory. Raj
and Ashby (27) have shown that the minimum volume required for void
stability is smaller in the case of spherical particles. The spherical oxides of the recrystallized single crystals of Cu-SiO$_2$ were
indeed found to be more efficient nucleation sites than the usual
lenticular oxides in the material oxidized in the polycrystalline
state (26).

The importance of microstructure in creep cavity nucleation
also has been demonstrated in a series of experiments (31,32) on
the nickel base superalloy, Astroloy, a material in which the slip
is highly planar. The Astroloy used in these studies contained
closely spaced M$_{23}$C$_6$ carbides elongated in the plane of the boundaries. Figure 7 shows a grain boundary in a specimen of Astroloy
crept at 750° C under a stress of 500 MPa to a total creep strain of
5.9%. The shadowed two-stage replica was taken from the surface of
a specimen which had been lightly electropolished. The trace of
the grain boundary shown in this micrograph is transverse to the
creep stress. Note the heavy slip traces which tend to impinge on
the carbides close to their ends. (A study of numerous micrographs
of such specimens showed that almost 80% of the slip traces hit a
boundary carbide at a point within 10% of its end.) This pronounced
strain localization occurs on all boundaries, independent of their
orientation with respect to the stress axis. However, the cavities
are largely confined to the transverse boundaries. About 70% of the
voids are associated with one or more slip traces. The increase in
the density of the heavy slip traces with creep strain is almost
identical to the dependence of void density on strain. A model has
been developed of the role of hard grain boundary particles in creep
cavitation which explains the experimental observations in a material
like Astroloy (highly planar slip, elongated grain boundary carbides
a few tenths of a micrometer in length). Both the difference in
elastic moduli between matrix and particle and their difference in
deformability lead to large concentrations of stress at the matrix/
carbide interface. Figure 8 shows a calculation, based on the equivalent inclusion model of Tanaka and Mura (33), of such stresses in
an ensemble of particle and matrix strained to a value ϵ^A. It can

Fig. 7. Shadowed two-stage replica of a sample of Astroloy crept at 750°C under a stress of 500 MPa to a strain of 5.9%. The surface of the sample was lightly electropolished. The grain boundary is transverse to the creep stress axis. Note the grain boundary voids (black spots with white shadow tails) located at the intersection of a heavy slip trace with the grain boundary. From (31).

be seen that, whether the strain is perpendicular or parallel to
the plane of the boundary, large shear stresses are set up close to
the ends of the carbides. These stresses can be big enough to initiate intense slip bands which produce a localized tensile stress
at the interface sufficient to form a microcrack (i.e., nucleate a
void). Since the stress concentrations around a particle are expected to fall off in a distance which scales with the particle
size, voids would be less likely to nucleate at very small particles.
Thus, at high temperatures as at low, particles become less damaging
as they decrease in size. During creep enhanced tensile normal
stresses acting across the interface next to a void nucleus drive
vacancies into the void in the case of particles on boundaries
transverse to the creep axis; on parallel boundaries large compressive stresses act to heal the voids (Fig. 8). As in nucleation,
the process of cavity growth is profoundly influenced by microstructure. The diffusion of vacancies appears to be responsible for
void growth in many cases. The segregation of certain solute and
impurity atoms to grain boundaries at high temperatures affects
grain boundary diffusivities (34). Void growth frequently is constrained (35) by the inability of neighboring grains to accommodate
the localized strain introduced by a cavitating boundary. Grain

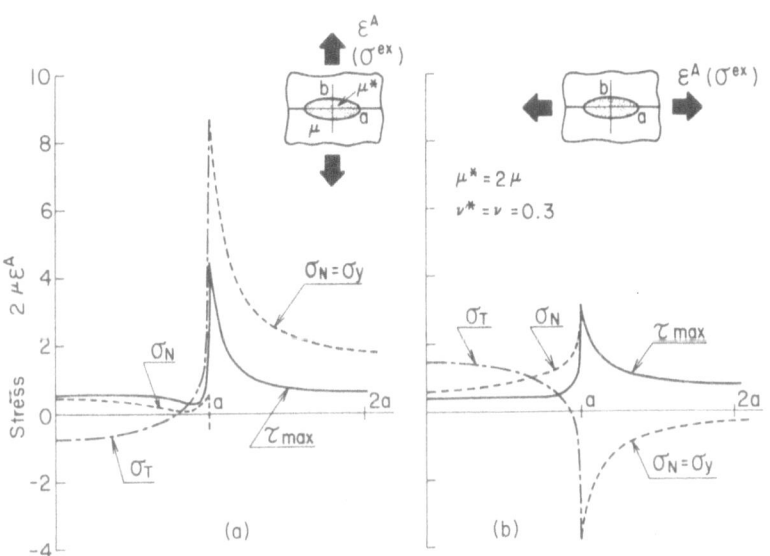

Fig. 8. The normal σ_N, tangential σ_T, and maximum shear τ_{max} stresses
which develop at the interface of an elastic inclusion in an elastic
matrix. The matrix/inclusion ensemble is pulled by an external stress
σ^{ex} which produces a strain ϵ^A far from the inclusion. From (31).

size and grain boundary morphology and particle population influence this rate of accommodation and thus the void growth rate.

Size, shape and spacing of the grain boundary particles; their interfacial energy; degree of slip planarity in the matrix; grain size--all of these microstructural features affect the rate of cavitation damage accumulation under creep conditions.

High Temperature Fatigue

The relationship between microstructure and fatigue-induced cavitation generally is similar to that under creep conditions. However, Figs. 5b and 6b show that cavitation levels under a high frequency cyclic stress do not appear to be related to edge-to-edge interparticle spacings but are strongly correlated with p/λ, the area fraction of the grain boundaries covered by particles. The lower the value of p/λ, the greater the stress concentration at the particles during grain boundary sliding. Thus it appears that grain boundary sliding is important in the process of fatigue cavitation, much less so in creep cavitation. This observation may be explained as follows. At high temperatures under static loading, the stress concentrations at particles caused by sliding boundaries are quickly relaxed by diffusional accommodation. But under cyclic stressing these stress concentrations never have a chance to subside and therefore continue to influence cavitation rates. It is not surprising that, under comparable applied stresses, nucleation rates are much greater under cyclic loading than under creep conditions (26,36).

Not all high temperature failure is intergranular, of course. For example, the ferritic stainless steels such as Fe9CrlMo are especially resistant to grain boundary damage at high temperatures. A modified version of this alloy, Fe9CrlMo containing small amounts of the strong carbide formers V and Nb, has been developed at ORNL (37,38) for power generation applications. The alloy must exhibit microstructural stability over many years even though subjected to creep and fatigue stresses at elevated temperatures. The Fe9CrlMo+V,Nb alloy, which is used in the tempered martensite state, derives its strength from fine MC carbides dispersed throughout the matrix, the lath substructure, and a high dislocation density. Figure 9 (39) shows the stress amplitudes which develop in the modified Fe9CrlMo as it is cycled to failure in fully reversed fatigue at 593° C through a total strain range of 1.2%. Three cycling patterns were used: continuous cycling, and cycling with a two minute hold at the peak tensile or peak compressive strain. It can be seen from Fig. 9 that the amplitudes decrease continuously throughout the testing, with the steel losing about 30% of its initial strength in a few hours to a few days. This cyclic softening is roughly the same as that produced by aging the steel under no load for 5000 h at 593° C. Another striking feature of these curves

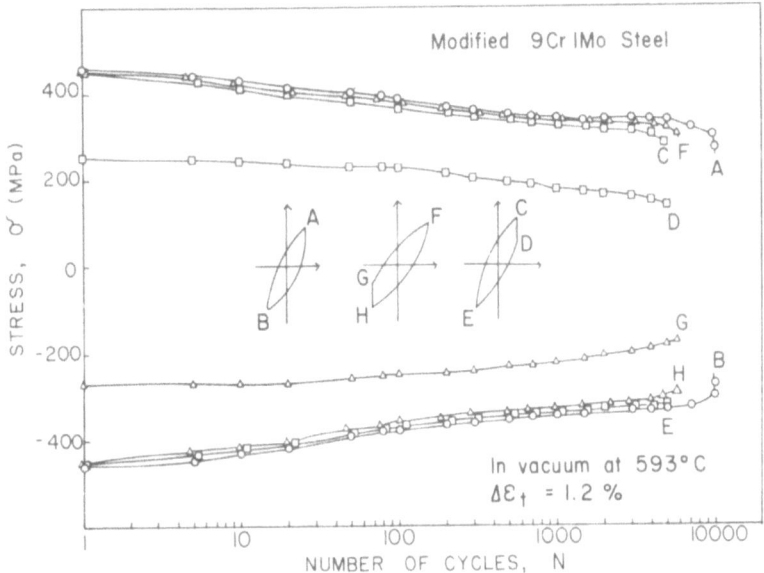

Fig. 9. Stress amplitudes as a function of number of cycles. Modified Fe9Cr1Mo tested in fully reversed fatigue through a total strain range of 1.2%, strain rate of 4×10^{-3}/s, 593°C, with and without 2 minute holds at peak strain. Note that the relaxed stress at the end of a hold time also is given.

is the large stress relaxation which occurs during the two minute hold. As might be expected, the mechanical behavior reflects the microstructural changes which are taking place. The fine intragranular VC and NbC particles disappear while boundary $M_{23}C_6$ carbides coarsen, the lath substructure present after normalization and tempering transforms to equiaxed subgrains and the dislocation density drops. These observations suggest that extensive research is needed into the factors affecting carbide precipitation and dissolution, with and without accompanying deformation, in these alloys if ferritic stainless steels are to be designed with the long term microstructural stability necessary for successful extended high temperature service.

The metallurgist wishes to understand phenomena such as fatigue and creep better not only to give a better basis for lifetime prediction but also to design more fatigue and creep resistant alloys. This discussion of low temperature fatigue, creep and high temperature fatigue illustrates just how difficult it is to achieve such improvement but hopefully it has presented some ideas for future research.

108

ACKNOWLEDGEMENT

    This research was carried out under the aegis of the Fatigue
Thrust Group of the Materials Research Center of Northwestern University, which is funded through the NSF-MRL program (Grant No. DMR-
8216972). Additional support came from AFOSR, NSF, DOE, ONR & Exxon.

REFERENCES

1. Hood, C. Institution of Civil Engineers (British), Minutes of
   Proceedings, vol. 2 (1842) 180.
2. Anton, D. L. and Fine, M. E. Fatigue Crack Evolution in Overaged
   Ni-14.4 at.% Al Alloy with Coherent Precipitates. Materials
   Science and Engineering 58 (1983) 132-142.
3. Fine, M. E. Fatigue Crack Initiation on Slip Bands in Proceedings
   of 6th Int. Conf. on Strength of Metals and Alloys (ICSMA6)
   R. C. Gifkins, ed., vol. 2 (Pergamon Press Ltd., 1982) 833-838.
4. Lin, M. R. Experimental and Theoretical Study of Fatigue Crack
   Initiation in Metals, Ph.D. thesis research in progress, Northwestern University, Dept. of Materials Science and Engineering,
   Evanston, Illinois.
5. Polmear, I. J., Bainbridge, I. F. and Glanvill, D. W. Cavity
   Formation During Metal Fatigue. Journal of the Australian
   Institute of Metals 7 (1962) 222.
6. Wilkov, M. A. University of Texas-Austin. Unpublished research.
7. Kung, C. Y. and Fine, M. E. Fatigue Crack Initiation and Microcrack Growth in 2024-T4 and 2124-T4 Aluminum Alloys. Metallurgical Transactions 10A (1979) 603-610.
8. Morris, W. L. Crack Closure Load Development for Surface Microcracks in Al 2219-T851. Metallurgical Transactions 8A (1977)
   1079-1085; A Comparison of Microcrack Closure Load Development
   for Stage I and II Cracking Events for Al 7075-T651. ibid. pp.
   1087-1093.
9. Thielen, P.N. Fatigue Studies on 4140 Quenched and Tempered
   Steel, Ph.D. thesis, Northwestern University, Evanston, Ill.,
   1975.
10. Lankford, J. The Growth of Small Fatigue Cracks in 7075-T6 Alumi-
    num. Fatigue of Engineering Materials Structures 5 (1982) 233.
11. Heubaum, F. Microstructural Effects of the Growth of Short Fa-
    tigue Cracks in HSLA Steels. Ph.D. research in progress, North-
    western University, Dept. of Materials Science & Engineering,
    Evanston, Illinois.
12. Masounave, J. and Baïlon, J. P. Effect of Grain Size on the
    Threshold Stress Intensity Factor in Fatigue of a Ferritic
    Steel. Scripta Metallurgica 10 (1976) 165-170.
13. Lin, G. M. and Fine, M. E. Effect of Grain Size and Cold Work
    in the Near Threshold Fatigue Crack Propagation Rate and Crack
    Closure in Iron. Scripta Metallurgica 16 (1982) 1249-1254.

14. Suresh, S. and Ritchie, R. O. A Geometric Model for Fatigue Crack Closure Induced by Fracture Surface Roughness. Metallurgical Transactions 13A (1982) 1627-1631.

15. Otsuka, A., Mori, K. and Miyata, T. The Condition of Fatigue Crack Growth in Mixed Mode Condition. Engineering Fracture Mechanics 7 (1975) 429-439.

16. Park, D. H. and Fine, M. E. Origin of Crack Closure in the Near-Threshold Fatigue Crack Propagation of Fe and Al-3% Mg. To appear in TMS-AIME Conf. Vol. "Concepts of Fatigue Crack Growth Threshold" (an Intl. Symp.) Philadelphia, Pa., October 2-5, 1983.

17. Minakawa, K., Matsuo, Y. and McEvily, A. J. The Influence of a Duplex Microstructure in Steels and Fatigue Crack Growth in the Near Threshold Region. Metallurgical Transactions 13A (1982) 439-445.

18. Horng, J. L. and Fine, M. E. Near-Threshold Fatigue Crack Propagation of HY80 and HY130 Steels. To appear in TMS-AIME Conf. Vol. "Concepts of Fatigue Crack Growth Threshold" (an Intl. Symp.) Philadelphia, Pa., October 2-5, 1983.

19. Horng, J. L. and Fine, M. E. Near Threshold Fatigue Crack Propagation Rates of Dual Phase Steels. Accepted for publication in Materials Science and Engineering.

20. Fine, M. E. Fatigue Resistance of Metals. Metallurgical Transactions 11A (1980) 365-379.

21. Fine, M. E. and Davidson, D. L. Quantitative Measurement of Energy Associated with a Moving Fatigue Crack in Fatigue Mechanisma: Advances in Quantitative Measurement of Physical Damage, ASTM STP 811, J. Lankford, D. L. Davidson, W. L. Morris and R. P. Wei, eds., Amer. Soc. for Testing & Materials, 1983, pp. 350-370.

22. Liaw, P. K., Kwun, S. I. and Fine, M. E. Plastic Work of Fatigue Crack Propagation in Steels and Aluminum Alloys. Metallurgical Transactions 12A (1981) 49-55.

23. Jeffries, Z. Effect of Temperature, Deformation, and Grain Size on the Mechanical Properties of Metals. Transactions AIME 60 (1919) 474-576.

24. Page, R. and Weertman, J. R. HVEM Observations of Grain Boundary Voids in High Purity Copper, Acta Metallurgica 29 (1981) 527-535.

25. Fleck, R. G., Taplin, D. M. R. and Beevers, C. J. An Investigation of the Nucleation of Creep Cavities by 1 MV Electron Microscopy, Acta Metallurgica 23 (1975) 415-424.

26. Chen, R. T. and Weertman, J. R. Grain Boundary Cavitation in Internally Oxidized Copper. Materials Science and Engineering 64 (1984) 15-25.

27. Raj, R. and Ashby, M. F. Intergranular Fracture at Elevated Temperature. Acta Metallurgica 23 (1975) 653-666.

28. Argon, A. S., Chen, I. W. and Lau, C. W. Intergranular Cavitation in Creep: Theory and Experiments, Creep-Fatigue-Environment Interactions, R. M. Pelloux and N. S. Stoloff, eds. (The Metallurgical Society--AIME, Warrendale, PA, 1980) 46-80.

29. Wood, S., Adamonics, A., Guha, A., Soffa, W. A. and Meier, G. H. Internal Oxidation of Dilute Cu-Ti Alloys. Metallurgical Transactions 6A (1975) 1793-1800.

30. Ashby, M. F. and Smith, G. C. Structures in Internally Oxidized Copper Alloys. Journal of the Institute of Metals 91 (1963) 182-187.

31. Shiozawa, K. and Weertman, J. R. The Nucleation of Grain Boundary Voids in a Nickel-Base Superalloy During High Temperature Creep. Scripta Metallurgica 16 (1982) 735-739.

32. Shiozawa, K. and Weertman, J. R. Studies of Nucleation Mechanisms and the Role of Residual Stresses in the Grain Boundary Cavitation of a Superalloy. Acta Metallurgica 31 (1983) 993-1004.

33. Tanaka, K. and Mura, T. A Theory of Fatigue Crack Initiation at Inclusions, Metallurgical Transactions 13A (1982) 117-123.

34. Padgett, R. A. and White, C. L. Retardation of Grain Boundary Self-Diffusion in Nickel Doped with Antimony and Tin. Scripta Metallurgica 18 (1984) 459-462.

35. Dyson, B. F. Constrained Cavity Growth, Its Use in Quantifying Recent Creep Fracture Results. Canadian Metallurgical Quarterly 18 (1979) 31-38.

36. Yang, M. S., Weertman, J. R. and Roth, M. Small Angle Neutron Scattering Studies of Microstructural Changes in Copper Deformed at Elevated Temperatures, Fifth Risø International Symposium on Metallurgy and Materials Science, 3-7 September 1984, Roskilde, Denmark. In press.

37. Sikka, V., Ward, C. T. and Thomas, K. C. Modified 9Cr-1Mo Steel-An Improved Alloy for Steam Generation Application, Ferritic Steels for High Temperature Applications, Ashok K. Khare, ed. (ASM, Metals Park, OH 44073, 1983) 65-84.

38. Booker, M. K., Sikka, V. K. and Booker, L. P. Comparison of the Mechanical Strength Properties of Several High-Chromium Ferritic Steels, Ferritic Steels for High-Temperature Applications, Ashok K. Khare, ed. (ASM, Metals Park, OH 44073, 1983) 257-273.

39. Kim, S., Chen, S.-H. and Weertman, J. R. To be published.

ETHANOL CRAZE GROWTH IN POLYCARBONATE

Maury Q. Falkoff and James A. Donovan

Mechanical Engineering Department
University of Massachusetts
Amherst, MA 01003

ABSTRACT

Ethanol craze growth kinetics in polycarbonate (PC) were studied using single edge notch (SEN) specimens over the temperature range 280-340K; the data were analyzed by fracture mechanics and kinetic rate theory.

The craze was modelled as a Dugdale zone and the composite stress intensity of the crack-craze controlled growth kinetics. The logarithm of the craze velocity increased linearly with the stress intensity for all test temperatures. However, this relation was not collinear for sequential growth steps. Below 337 K the threshold stress intensity appears to be athermal; but above 337 K the model was not applicable.

The Dugdale craze stress decreased linearly with temperature, and extrapolated to zero at 348 K, an estimate of the glass transition temperature (T_g) of the plasticized craze as affected by the environment and local stress. The activation enthalpy for craze growth increased with craze driving force; with no mechanical driving force the activation enthalpy was 4 Kcal/mole. The activation area increased with temperature, decreased with mechanical driving force and was small compared to molecular dimensions. Plasticization (decrease of T_g) by the ethanol appears to be the primary mechanism of ethanol craze growth in PC.

INTRODUCTION

Polymers are being used increasingly for structural applications because of their ease of fabrication, relatively low cost, low density, low electrical resistivity, and chemical inertness. Despite such advantages degradation of mechanical properties occurs in agressive environments and may lead to rapid failure. Therefore, thermal characterization of environment assisted deformation processes in polymeric materials is imperative.

Crazing, a local deformation process, creates crack like defects, but the craze surfaces are connected by fibrils that support a load. Craze growth is accelerated in some environments and usually precedes crack growth in glassy polymers. The objective of this research was to characterize the mechanical driving force and temperature dependence of ethanol crazing in PC. Matsushige et al.(1) suggested that the criterion for environmental crazing is when the largest principal stress exceeds a critical value. While the maximum principal stress criteria is useful for predicting behavior in smooth specimens it is of little help in evaluating the role of stress concentrators on craze growth. For this reason other researchers have used linear elastic fracture mechanics to describe craze growth, in terms of the stress intensity of the starting flaw.

Kramer et al. (2) have measured craze opening displacement profiles with double exposure holographic interferometry and calculated craze stress profiles for a variety of organic polymer solvent combinations. Their results show that the craze stress is approximately uniform, except for a large stress peak near the craze tip, in general agreement with the Dugdale model. Gerberich (3) proposed that there is a plastic zone at the tip of a growing craze due to the applied stress intensity, and that there is a threshold stress intensity which corresponds to a minimum plastic zone size at the craze tip.

Williams and Marshall (4) tested PMMA and PC SEN and surface notched specimens in ethanol and found that in order for crazing to occur the stress intensity of the starter crack had to exceed a minimum value, and that in order for crazing to lead to fracture it had to exceed a higher value. For intermediate values of stress intensity craze growth eventually arrested.

Priori et al. (5) studied PC crazing in specimens with a strain gradient in normal hydrocarbons ranging from n-hexane to n-dodecane. They found that the activation energies for craze growth decreased linearly with the square of the difference between the solubility parameter of PC and the crazing agent, from 41,360 cal/mole with n-hexane to 28,130 cal/mole with n-dodecane. Their experiments did not account for stress relaxation, and the strain energy in their specimens was dissipated by the nucleation and propagation of many crazes. For these reasons their activation analysis does not describe the energetics of crazing accurately.

Modified Dugdale Model for a Craze

The Dugdale model (6) estimates the size of the plastic zone ahead of a crack tip in a thin sheet and has been used to characterize craze length in polymers (7). The stress intensity relationship for growth of a craze was taken as

$$K_{ck} + K_{cz} > K_{th} \tag{1}$$

where K_{ck} is the stress intensity for a crack of length a_0 (See Fig 1), K_{cz} is the stress intensity for a craze of length $a-a_0$ and K_{th} is the threshold stress intensity below which the craze will not grow (3). Inclusion of K_{th} is a modification of the original Dugdale model. K_{ck} is obtained by the standard methods of linear elastic fracture mechanics. K_{cz} was shown by Williams (8) to be

$$K_{cz} = -2\sigma_{cz}\sqrt{a/\pi}\; \cos^{-1}\left(\frac{a_0}{a}\right) \tag{2}$$

where σ_{cz} is a uniform stress acting over the length of the craze. Note that K_{cz} is negative thus shielding the crack tip and reducing the net stress intensity. Therefore, the effective stress intensity is

$$K = K_{ck} + K_{cz}. \tag{3}$$

For a single edge notch specimen the effective composite stress intensity of the crack-craze is

$$K = \sigma_\infty \sqrt{a}\; Y(a/w) - 2\sigma_{cz}\sqrt{a}\cos^{-1}(a_0/a)\; Y(a/w)/\pi \tag{4}$$

where σ_∞ is the applied tensile stress, $Y(a/w)$ is the appropriate geometry factor and w is the specimen width.

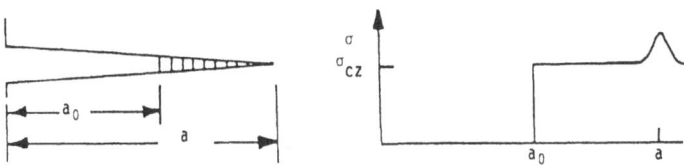

Figure 1. The crack-craze system and craze stress profile.

Note that Equation 4 uses the Dugdale model only in the sense that it assumes a uniform craze stress acts along the craze length $(a-a_0)$. Equation 4 does not restrict the length of the craze in relation to the length of the starting crack except for the limitations imposed by ASTM standards for valid ranges of a/w.

Thermally Activated Craze Growth

Reaction rate theory applied to craze growth by analogy with crack growth (9) assumes the temperature dependence of craze velocity to be

$$V_{cz} = V_0 \ (J,T,s) \ exp- \ \frac{\Delta G(J,T,s)}{kT} \tag{5}$$

where V_0 is the maximum velocity and is a function of J, the craze driving force per unit width, T the absolute temperature and s a structure characterizing parameter, ΔG is the change in Gibbs free energy for the process and k is Boltzman's constant. The craze driving force may have an athermal component J_{th}, therefore, $J*$ the effective craze driving force for overcoming thermal obstacles is

$$J* = J - J_{th} \tag{6}$$

where J is the applied craze driving force and is related to the stress intensity; for linear elastic behavior in plane stress $J = K^2/E$.

The activation energy can be shown to be

$$\Delta H = -k \ \frac{\partial \ lnV}{\partial \ 1/T} \bigg|_{J*} \tag{7}$$

and the activation area is

$$A = kT \ \frac{\partial \ lnV}{\partial \ J*} \bigg|_T \tag{8}$$

These equations are valid as long as the nature and concentration of thermal obstacles in the craze tip region remain constant, that is as long as s remains constant. This analysis also assumes that craze growth is thermodynamically reversible, and that the forward activation process predominates, a realistic assumption.

EXPERIMENTAL PROCEDURE

Stress relaxation due to growth of a single craze in SEN polycarbonate (Lexan, G.E. Co.) specimens immersed in ethanol was determined. Also craze length was measured with a calibrated optical microscope, so that both the craze length and applied load were known as a function of time and temperature between 280-340K. The specimens were machined to final dimensions (250 x 51 x 9.2 mm,) without cutting fluids or lubricants. Starter cracks were made by slowly driving a new utility razor blade into the notch tip.

All specimens were vacuum annealed at 393 K for 24 hours and oven cooled to relieve residual stresses and heal microflaws from machining and pre-cracking. Differential scanning calorimetry showed that the annealing did not densify, order or crystallize the PC, and that its T_g was 427 K.

An environmental chamber with a clear glass front contained the ethanol and a small heat exchanger, which controlled the ethanol temperature within ± 0.5 K.

During the stress relaxation test at fixed displacement the load relaxed due to the growth of a single craze. Eventually, the load decreased to a constant value and the craze stopped growing. This condition of load and craze size was taken as the threshold or equilibrium condition and was the basis for calculating K_{th}. Several craze growth steps were studied on a single specimen as shown in Figure 2; the equilibrium conditions correspond to points A, B and C.

For two different steps I and J, it was assumed that Kth(I)=Kth(J), and therefore, the craze stress, σ_{cz} could be calculated from

$$\sigma_{cz} = \frac{\pi}{2} \frac{[\sigma\infty(J)\sqrt{a(J)}Y(J)-\sigma\infty(I)\sqrt{a(I)}Y(I)]}{[\sqrt{a(J)}\cos^{-1}(a_0/a(J))Y(J)-\sqrt{a(I)}\cos^{-1}(a_0/a(I))Y(I)]}. \qquad (9)$$

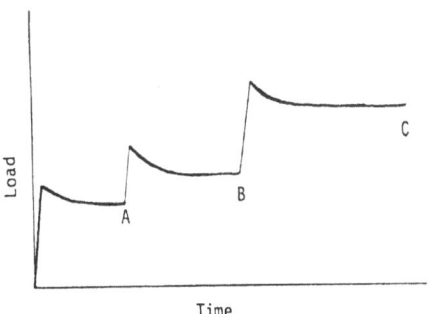

Figure 2. A typical load vs time curve for a 3 step experiment.

The mean value of σ_{cz} was then used in Equation 4 to calculate the threshold stress intensity.

RESULTS

The ℓnV vs. K plots were linear (Fig. 3a), with high correlation coefficients; however, the ℓnV vs. K curves for sequential load relaxation tests were not collinear, but consistently shifted to the right, suggesting a small systematic error in the stress intensity as a function of craze length. However, useful and consistent results were derived from these data, as will be shown.

The values of K_{th} for seven test temperatures between 280 and 337 K are shown in Fig. 3b, the variation in K_{th} for different load relaxation steps is shown. The values all have the magnitude of 10^{-1} MN/m$^{1.5}$, and agree within experimental error. The consistency of the calculated values of K_{th}, suggests that K_{th} is athermal at temperatures below 337 K. However, at 337 K the calculated K_{th} was negative. There is no physical significance to this value; therefore the proposed model appears inappropriate for $T \geq 337$ K.

σ_{cz} decreased linearly with temperature and extrapolated to zero at 348 K (Fig. 4). This is the glass transition temperature of the craze as influenced by the local stress and ethanol environment. This also explains the unusual craze growth behavior at 337 K; since this temperature is near the apparent Tg of the craze the mechanism of craze growth is apparently changing rapidly.

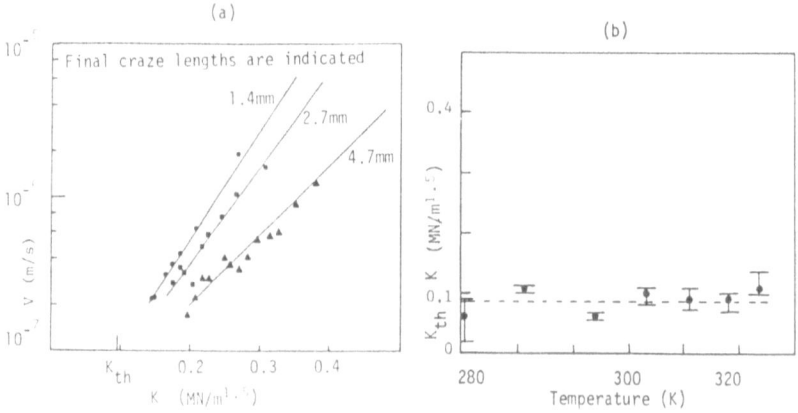

Figure 3 (a) Craze velocity as a function of stress intensity at 313 K, (b) The mean threshold stress intensity as a function of temperature.

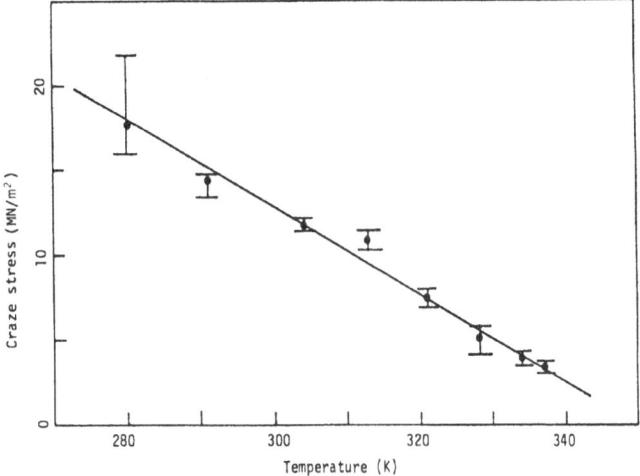

Figure 4. The craze stress as a function of temperature.

Figure 5a shows temperature dependence of V at three levels of J*. The activation enthalpy, from the slopes, are 5.9, 8.5 and 11.6 Kcal/mole for craze driving forces of 23,48, and 77 N/m, respectively. The scatter in ℓnV at J* = 48N/m as a function of temperature is shown in Fig. 5. The scatter at other values of craze driving force was similar. The activation enthalpy increased with driving force, and can be given as $\Delta H = 4 + 0.1J$ in Kcal/mole. The activation area was calculated using the approximation that J* = J, which is a reasonable approximation when J is well above threshold conditions ($J_{th} \approx 3$ N/m).

Figure 6 shows the dependence of the activation area at five temperatures. The values of A range between 5 x 10^{-3} and 5 x 10^{-2} $\mathring{A}^2$, values which are small compared to molecular dimensions.

DISCUSSION

The linear relation of ℓnV with the composite stress intensity due to the crack and craze suggests that J is the appropriate driving force of craze growth. This is also supported by craze growth measurements in compact tension specimens that showed the same ℓnV vs K relations as SEN specimens(11).

The lack of collinearity in ℓnV as a function of K during sequential steps, however, indicates that the parameters in the stress intensity relation need refinement. Several possible

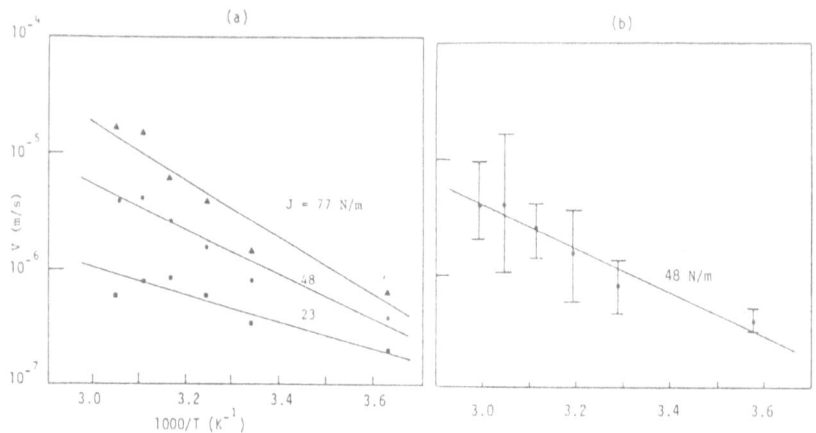

Figure 5. (a) The craze velocity as a function of 1/T. (b) The scatter in (a) at J = 48 N/m.

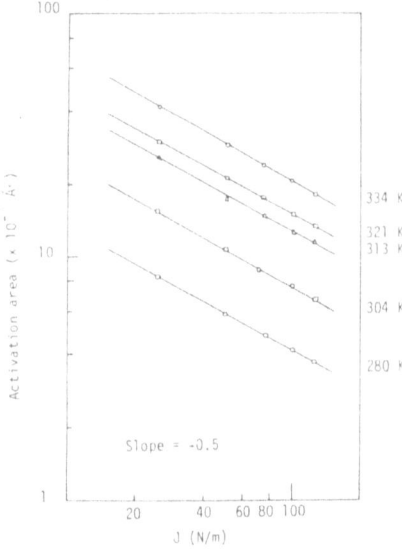

Figure 6. The activation area as a function of craze driving force at different test temperatures. The craze lengths were between 2.0 and 3.2 mm.

errors may be responsible: 1) No correction was made for the curvature of the craze profile, 2) The assumption of a constant craze stress may not be accurate due to relaxation of the fibril stress, and 3) The adsorption of the ethanol into the PC may have produced a swelling stress. Clarification of these points will be the basis of future work.

Accurate determination of threshold stress intensity is always a problem, but the data suggest that it is an athermal material property of the ethanol – PC system, albeit small. The fact that it is small is not unusual for polymers.

The role of the ethanol is primarily as a plasticizer. This conclusion is based on the temperature dependence of the craze stress which decreases linearly to zero at 348 K, which would be the T_g of the craze. Also, the negative value of K_{th} at 337 K suggests that the mechanism of craze growth is in transition so that the model proposed is appropriate only at temperatures somewhat below T_g. In the temperature range near T_g the kinetics of the molecular processes are changing very rapidly as manifested in the mechanical behavior of the craze. The visual appearance and growth characteristics of the craze also changed near the T_g of the craze.

That the activation enthalpy increased with craze driving force implies that the activation entropy increases since the Gibbs free energy must decrease. This increase in entropy is related to the structural and volume changes which occur due to the formation of the craze fibrils and voids.

The decrease of the activation area with driving force was expected, but its small size as compared to molecular dimensions requires additional insight to be understood. Activation areas of similar magnitude were obtained by other investigators for fracture of glassy polymers (10).

Identification of the rate controlling mechanism of craze growth is uncertain, but based on this data the best speculation is that diffusion of the ethanol into the craze tip is the rate controlling process. This is based on the similarity of the activation enthalpy at zero driving force with that for diffusion of organic liquids in glassy polymers (12).

CONCLUSIONS

The mechanical driving force that controls craze growth is related to the composite stress intensity due to the crack and craze. This description requires some refinement in that the relation of craze velocity to driving force for sequential growth steps were not collinear. However, the data derived from the stress relaxation tests show that the ethanol lowers the T_g of the craze from 427 to 348 K and that the activation enthalpy and area for craze growth are functions of the driving force. The ethanol functions as a plasticizer and the rate controlling process may be the diffusion of the ethanol in the PC.

ACKNOWLEDGEMENT

We are greatful for the support of this work by the Materials Research Laboratory at the University of Massachusetts funded by the U.S. National Science Foundation. Also, we would like to thank General Electric Co. for the polycarbonate.

REFERENCES:

1) Matsushige, K., S.V. Radcliffe and E.J. Baer, J. Mater. Sci, 10 (1973) 833.
2) Kramer, E.J., Chapter 1, Advances in Polymer Science, 50, Springer Verlag, 1983.
3) Gerberich, W.W., Int. J. Fracture, 13, (1977) 535.
4) Williams, J.G., and G.P. Marshall, Proc. Royal Soc., A-342 55, (1975).
5) Priori, A., L. Nicolais, and A.T. DiBenedetto, J. Mater, Sci., 18, (1983) 1466.
6) Dugdale, D.S., J. Mech. and Phys. of Solids, 8, (1960) 100.
7) Graham, I.D., J.G. Williams and E.L. Zichy, Polymer 17 (1976) 439.
8) Williams, J.G., Stress Analysis of Polymers, London, Longmans, (1973).
9) Pollet, J.C., and S.J. Burns, Int. J. of Fracture, 14 (1977) 667
10) Pollet, J.C., and S.J. Burns, Ibid, 775.
11) Lin, S.S., M.S. Thesis in Mechanical Engineering, University of Massachusetts (1983).
12) Diffusion in Polymers, Ed. J. Crank, Academic Press (1968) 235.

EFFECTS OF MICROSTRUCTURE, COLD WORK AND ANISOTROPY ON CREEP CRACK GROWTH IN CARBON MANGANESE STEEL AT 360°C

D.J. Gooch

Technology Planning and Research Division
Central Electricity Generating Board
Central Electricity Research Laboratories
Kelvin Avenue, Leatherhead, Surrey KT22 7SE, England

ABSTRACT

Creep crack growth in carbon manganese steel at 360°C has been observed and is strongly dependent on degree of cold work, microstructure and orientation. The effect of 15% cold work is to increase crack velocities dramatically when the crack plane is parallel to the pre-strain direction. In the transverse direction the effect is less severe, the difference being attributed to variations in creep strength and inclusion distribution. Crack growth susceptibility in non-pre-strained material has been found to be a function of austenitizing temperature and microstructure, higher austenitizing temperatures resulting in increased crack velocities. Ferrite/pearlite and bainitic microstructures are more susceptible to crack growth than tempered martensite.

1   INTRODUCTION

Pressure containing components designed for operation at approximately 360°C in the electricity supply industry are frequently manufactured from carbon manganese steels. At this temperature the creep resistance is such that design may be based upon tensile properties and it is generally assumed that creep processes are insignificant. However in recent years there have been a number of worldwide instances of failure of cold formed carbon manganese steels operating in this temperature regime and attention is now focussed on creep as a likely mechanism.

This Paper describes an experimental programme of creep crack growth tests at 360°C on carbon manganese steel using ½T compact tension specimens. The parameters investigated include degree of cold work, orientation and variations in microstructure achieved by different austenitizing temperatures and cooling rates.

2   EXPERIMENTAL

Details of the steels investigated are given in Table 1.

HFS35 had been in service for 65,000 hours at 360°C and it was considered necessary to test it in the renormalized condition as well. Two renormalization treatments were adopted. Firstly, one hour at 900°C followed by an air cool giving a ferrite/pearlite structure of 15 µm grain diameter; compared with ∿25 µm for the ex-service condition. Secondly, one hour at 1100°C followed by a controlled cool at ∿400°C/hour giving a grain size of ∿30 µm. HS and LS were experimental hot rolled plates with high and low sulphur levels and a grain size of ∿10 µm.

Large rectangular tensile specimens (160 × 70 × 14 mm gauge) were gridded and strained at room temperature to average strains of up to 25%. Compact tension (CT) test pieces (W:25 mm, B:12.5 mm) were machined from these with the crack plane either (a) parallel to the tensile axis and perpendicular to the rolling plane (0°) or (b) perpendicular to the tensile axis and perpendicular to the rolling plane (90°C). Specimens of HFS35 without pre-strain were also machined in the 0° orientation from blanks which had been austenitized at temperatures between 900°C and 1300°C followed by different cooling rates. The final 3 mm of the notch were formed by spark erosion giving an a/W of 0.4 and a root radius of 0.15 mm. Creep crack growth and uniaxial creep specimens (gauge length 25 mm, diameter 5.65 mm) were tested under constant load in air at 360°C.

Table 1

Materials Details

Designation	Form	Hardness VPN	C	Mn	Si	S	Nb
HFS35	Pipe	190	0.30	1.07	0.19	0.033	–
HFS35 (HS)	Plate	192	0.30	0.99	0.30	0.045	–
HFS35 (LS)	Plate	187	0.30	0.94	0.20	0.004	–
490 Nb	Pipe	166	0.24	1.15	0.37	0.016	0.11

3   RESULTS

3.1 Effect of Pre-Strain

Endurance data at 360°C for CT specimens of the HFS35, HS and LS steels are plotted as a function of the plane stress reference stress in Fig. 1 and Fig. 2 together with the extrapolated ISO uniaxial rupture data for carbon manganese steels.

When plotted against the plane stress reference stress the endurance data for the renormalized HFS35 (Fig. 1) fall between the mean and -20% stress ISO data lines for carbon manganese steels. However the endurances for the pre-strained specimens fall below the -20% stress line, the shortfall increasing with the degree of pre-strain. The single test on the renormalized (900°C) material pre-strained to 15% gave a result virtually the same as the similarly pre-strained renormalized (1100°C) sample. This may be due to the opposing effects of the higher hardness but finer grain size of the former. The endurance of the unstrained sample at 300 MPa was $\sim$12,500 hours compared with $\sim$20,000 hours for the 1100°C renormalized material. In this case the higher hardness of the former, resulting from its faster cooling rate, appears to be out-weighing the finer grain size.

The uniaxial rupture data for the 15% pre-strained HFS35 lie above the ISO + 20% line such that the endurances of the correspond-ing crack growth samples are approximately one hundred times shorter than would be predicted by application of the reference stress to the uniaxial data.

The experimental steels with high and low sulphur levels, (HS, LS) pre-strained to 15%, showed endurances nearly 100 times shorter than the renormalized HFS35 with similar pre-strains (Fig. 2). However the hardnesses of the experimental steels were also approximately 40 points higher than the pipe steels after pre-straining. The endurances of the low sulphur steel were marginally higher than for the high sulphur but the difference could equally be related to the slightly higher hardness of the latter.

Tests on the niobium stabilized 490 Nb steel showed similar trends to the HFS35 grade except that the pre-strain effects were less severe. This was attributed to the lower creep resistance of the finer grained, lower carbon level 490 Nb (Fig. 3).

Increases in degree of pre-strain are accompanied by increases in hardness and one of the most striking features of the current work was the dramatic empirical relationship between crack sus-ceptibility and hardness. This is illustrated in Fig. 4 for the HFS35, HS and LS steels.

124

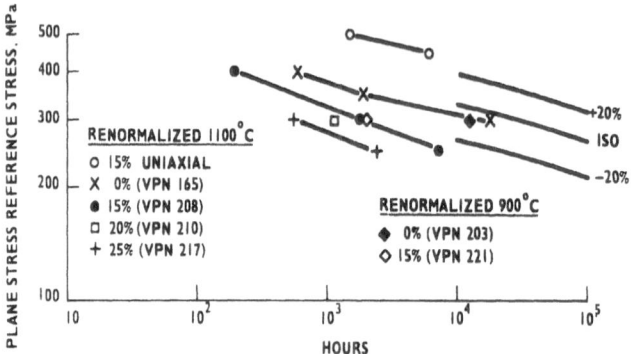

Fig. 1   Rupture Endurance for Renormalized HFS35

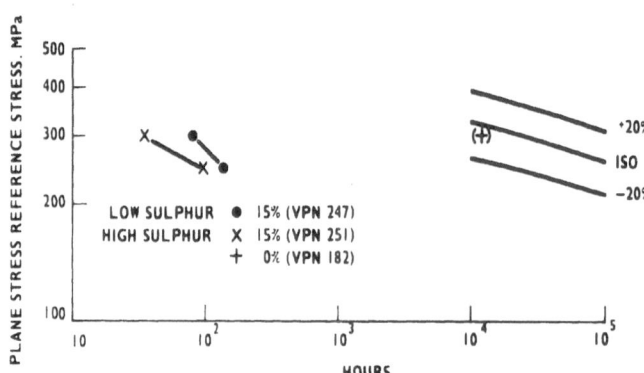

Fig. 2   Rupture Endurances for High and Low Sulphur HFS35

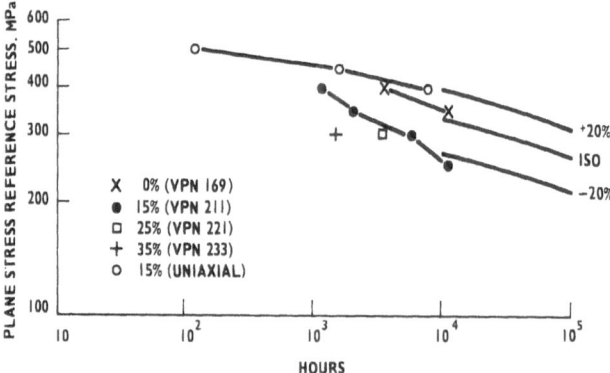

Fig. 3   Rupture Endurances for 490 Nb

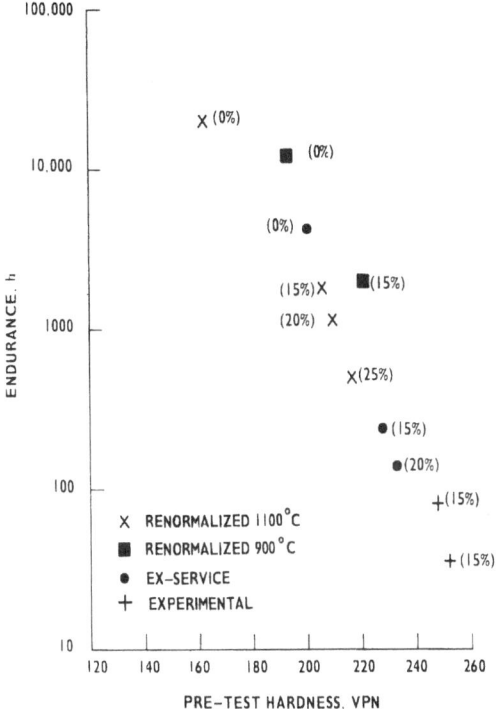

Fig. 4   Effect of Hardness on Endurance of HFS35; 300 MPa

3.2   Effect of Orientation

In the case of the 490 Nb steel (Fig. 5) the 90° specimens
showed greater resistance to crack growth than the corresponding 0°
samples, endurances increasing by a factor ranging from 1.25 without
pre-strain to 4.67 with 25% pre-strain. The effect of orientation
was thus comparable with the effect of pre-strain such that the 90°
sample with 15% pre-strain actually failed in a marginally longer
time than the 0° sample without pre-strain.

The magnitude of the orientation effect was similar for the
HFS35 grade, the endurance factor ranging from 1.25 for the grain
coarsened sample without pre-strain to 7.27 for the 1100°C renormalized sample with 15% pre-strain (Fig. 6). However, unlike the
490 Nb grade, the orientation effect was not sufficient to raise the
endurances of 90°, 15% pre-strain samples above those of 0°, 0%
pre-strain specimens. This was because the effect of pre-strain on
the endurance of the HFS35 steel was greater than on the specific
cast of 490 Nb tested in the present work. It should be noted that

126

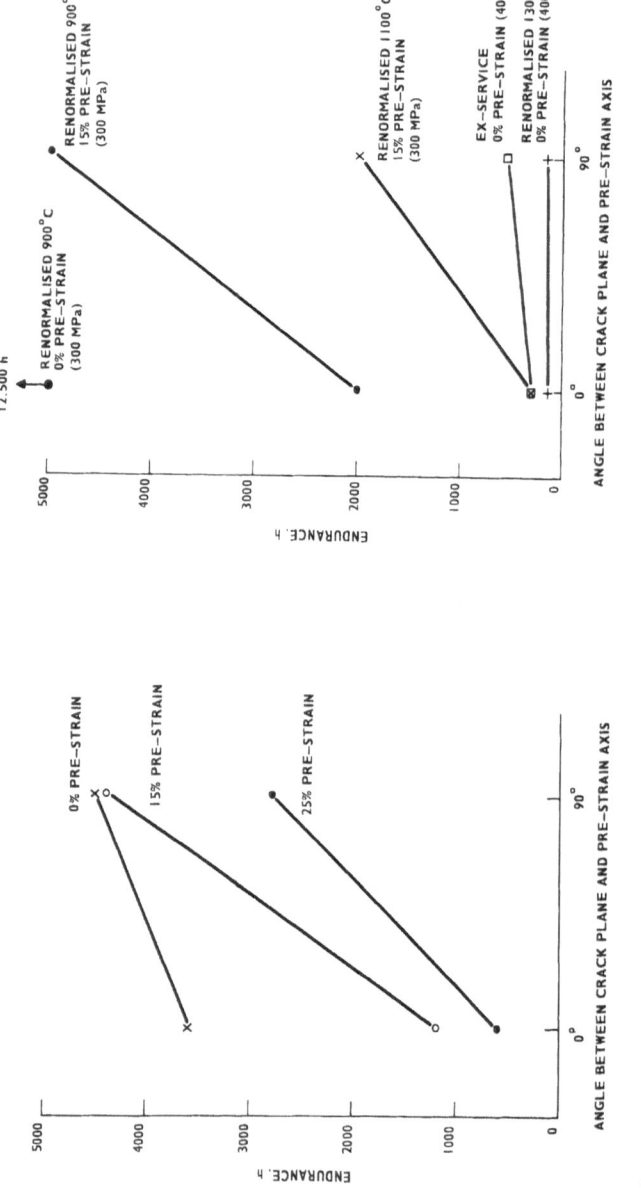

Fig. 5 Effect of Orientation on Endurance
of 490 Nb, 400 MPa

Fig. 6 Effect of Orientation on Endurance of
HFS35

the effect of pre-strain is cast-sensitive and that other casts of 490 Nb may show a greater pre-strain effect (1).

In the absence of pre-strain minimum creep rates of the 490 Nb steel were virtually indistinguishable although rupture elongations and reductions in area were higher for the axial orientation (Table 2). The higher ductility was also the cause of the longer rupture life for the 400 MPa axial sample, the increase being due to an extended period of tertiary creep.

After 15% pre-strain there was a clear separation of the axial and transverse properties, the minimum creep rates of the former being approximately twice as high as the latter. The increase in creep strength caused by pre-strain is considered to be a con-tributory factor to the effect of cold work in promoting suscept-ibility to creep crack growth.

Over the strain rate range studied the stress indice of the minimum creep rate varied from ∿23 at 10^{-3}/h to ∿12 at 10^{-5}/h, this being consistent with other work on the creep of C-Mn steel at this temperature (2).

Table 2

Effect of Orientation on Creep Properties of 490 Nb

Stress MPa	Orientation	Pre-Strain %	Min. Creep Rate h^{-1}	Elong. %	Rupture Life, h
450	Transverse	0	2.0×10^{-3}	33	71
400	"	0	1.3×10^{-4}	30	775
350	"	0	2.5×10^{-5}		Continuing
450	Axial	0	2.3×10^{-3}	48	60
400	"	0	1.5×10^{-4}	44	1150
350	"	0			Continuing
500	Transverse	15	4.2×10^{-4}	22	120
450	"	15	4.9×10^{-5}	30	1581
400	"	15	1.0×10^{-5}	24	7940
500	Axial	15	1.4×10^{-3}	65	65
450	"	15	9.5×10^{-5}	33	930
400	"	15	1.8×10^{-5}		Continuing

128

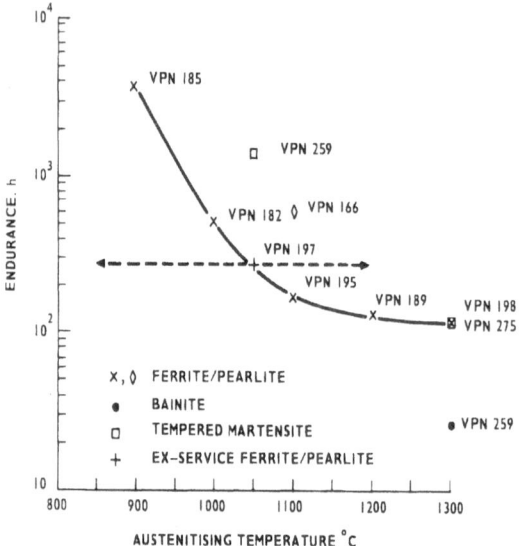

Fig. 7 Effect of Microstructure on Endurance of HFS35, 400 MPa

3.3 Effect of Microstructure

The endurances of the compact tension specimens as a function of austenitizing temperature are shown in Fig. 7.

Considering firstly the air cooled specimens it is clear that there is marked dependence of endurance on austenitizing temperature, particularly between 900°C and 1100°C when the endurance falls by a factor of ∿20. Over the same range the hardness only increases by ∿10 VPN. At 1100°C the microstructure has changed from a uniformly fine grained ferrite/pearlite structure to a duplex coarse pearlite/fine prior austenite boundary ferrite structure. However it is improbable that the decreased endurance is due to this change alone since the 900°C and 1000°C structures were indistinguishable both in terms of grain size, hardness and transformation product morphology but still showed a factor of 7 difference in endurance. The decreasing endurance was accompanied by a corresponding increase in crack aspect ratio reflecting the increasingly brittle nature of the crack propagation mechanism. Austenitizing temperatures above 1100°C had little effect on either endurance or crack aspect ratio.

Oil quenching from 1300°C produced a coarse grained bainitic structure of hardness 259 VPN which had an endurance a factor of 4.5 lower than the equivalent air cooled sample of hardness 198 VPN and a correspondingly higher crack aspect ratio. However the tempered martensite (275 VPN) formed by water quenching from 1300°C had an endurance similar to the softer air cooled sample. Similarly a

tempered martensite (259 VPN) formed by water quenching from 1050°C
endured five times longer than expected for the equivalent air
cooled sample.

Slow cooling at 400°C/hour from 1100°C resulted in a uniform
ferrite/pearlite structure of 166 VPN which lasted 3.5 times as long
as the harder air cooled sample and had a lower aspect ratio
reflecting its increased ductility.

Finally the ex-service sample, which was thought to have entered
service in the hot worked condition, exhibited an endurance corresponding to that of an air cooled structure after austenitization at
1050°C.

4. DISCUSSION

Pre-strain may influence crack susceptibility by two mechanisms.
Firstly there is a rapid increase in immobile dislocation density
which raises the hardness, but more importantly, the creep resistance.
Relaxation of crack tip stresses will thus be retarded favouring
cavitation or microcracking processes controlled by maximum principal
stress or hydrostatic stress components (3). Secondly pre-strain
generates internal stress fields by the formation of dislocation
pile-ups at grain boundaries, carbides and inclusions (4). Dyson (5)
has also shown that internal residual stress fields can be generated
by imbalances in the transmission of slip bands across boundaries.
These have been cited as the driving force for cavity growth from
microcracks during the annealing of pre-strained Nimonic 80A.

In the present study cracks were primarily intergranular with
respect to the ferrite grains and pearlite colonies although
decohesion of inclusion/matrix boundaries also occurred. Even if
microcracks were not formed directly by the room temperature pre-
strain, residual stresses could raise the effective grain boundary
energy and hence increase the ease of cavity or microcrack nucleation
by grain boundary sliding or shear during creep. Subsequent enlarge-
ment would then occur by a continuum mechanism (6) promoted by the
triaxiality of the stress state.

The orientation dependence of the crack growth susceptibility
arises out of a combination of the creep strength anisotropy observed
and the presence of inclusion stringers lying parallel to the pre-
strain direction. The inclusion stringers are not considered to be
the dominant influence on crack growth in the HFS35, HS and LS steels
and this is confirmed by Fig. 6 where the 90° pre-strained HFS35
remains less crack resistant than the 0° sample without pre-strain.
However in the lower creep strength and hence higher crack resistance
490 Nb steel the inclusions may become more significant thus accounting, together with the creep strength anisotropy, for the superior

crack resistance of the 90°, 15% pre-strained 490 Nb (Fig. 5).

Variations in creep resistance are also considered to be a major cause of the effect of austenitizing temperature on crack growth (Fig. 7). The creep strength of C-Mn steel is increased by the presence of free nitrogen and it has been suggested (7) that this arises from a clustering effect of manganese and nitrogen atoms in solid solution. The HFS35 steel used in this study was aluminium treated and in such steels nitrogen combines with aluminium to form AℓN during slow cooling. However as the austenitizing temperature is raised AℓN is taken into solution and may not be completely reprecipitated on cooling thus resulting in increased creep resistance. The superior crack resistance of the martensitic structures may also be influenced by creep strength effects since at the tempering temperature of 600°C free nitrogen would be expected to recombine with aluminium.

The major significance of this study lies in the demonstration that rapid, low displacement, intergranular creep crack growth may occur in particular metallurgical conditions of C-Mn steels at temperatures where creep was previously considered unimportant. The potentially susceptible microstructures are those which occur as a consequence of cold work or welding without stress relief and this should be taken into account during component design and manufacture.

5. ACKNOWLEDGEMENT

This work was performed at the Central Electricity Research Laboratories and the paper is published by permission of the Central Electricity Generating Board.

6. REFERENCES

1. Gooch, D.J. Mater. Sci. Eng. To be published.
2. Neumann, P. Unpublished Work.
3. Cane, B.J. Metal Science, 1981, vol. 15, 302.
4. Stroh, A.N. Proc. Roy. Soc., 1954, vol. 223A, 404.
5. Dyson, B.F. Canadian Metallurgical Quarterly, 1974, vol. 13, 237.
6. Hancock, J.W. Metal Science, 1976, vol. 10, 319.
7. Hopkin, L.M.T. JISI, 1965, vol. 203, 583.

STRESS CORROSION LIFE EXPECTANCY TOPOGRAPHS FOR THE DETERMINATION
OF DESIGN AND INSPECTION DATA

K. Krausz and A.S. Krausz[+]

[+]Department of Mechanical Engineering
University of Ottawa
770 King Edward Ave.
Ottawa KlN 6N5 Ontario, Canada

ABSTRACT

A theoretically rigorous derivation of lifetime determination
for stress corrosion cracking is presented. It is shown that
under certain conditions the crack velocity relation, obtained
from fracture kinetics analysis of thermally activated processes,
may be integrated directly and represented in the temperature —
time — crack driving force coordinate system. The step-by-step
construction of a simplified three-dimensional diagram is shown for
the three regions of SCC. The expected shape of the surface is
developed in detail for Region I. It is pointed out that for design, test, and reliability engineering purposes the conversion of
a mathematical expression into a three-dimensional diagram is
greatly advantageous.

INTRODUCTION

For the designer of structures and components environmental
effects are of great concern: stress corrosion cracking is one of
the major causes of material failure. The complex processes of
stress corrosion are studied extensively and much information is
available from the analyses of the experimental results. These
results, however, are not readily available to the design engineer.
The life expectancy of the selected material under the operating
conditions is an essential information for design purposes; it also
aids the test engineer to determine the inspection intervals. Thus,
the construction of life expectancy diagram, derived rigorously from
SCC studies, is particularly useful for both engineers.

The lifetime depends on the crack velocity, which in turn is
the function of the crack driving force (or stress intensity), the

temperature, the environment, and the material. It will be shown
that, under appropriate conditions, life expectancy calculations
may be carried out for the three regions of SCC by integrating
simple exponential equations. The calculations are based on physi-
cally rigorous relations that were developed for the study of thermally activated fracture of linear elastic solids [1]: the topographs will be constructed in the coordinate system of crack driving
force (G_i) — life-time (t_c) — temperature (T).

THE FRACTURE KINETICS BASE OF SCC

It is well established that stress corrosion is controlled by
a system of thermally activated processes [2-6]. The study of
thermal activation has led to fracture kinetics analyses that incorporate rate theory and fracture mechanics without the aid of any
empirical relations. It has been shown [7] that in SCC the crack
propagation velocity v can be described rigorously and precisely
by

$$v = \lambda k, \tag{1}$$

where λ is the atomic distance travelled by the crack during an
activation and k is the rate constant, defined by the rate theory
[8] and expressed for any of the three regions of SCC (Fig. 1) as

$$k = \frac{kT}{h} \exp\left(-\frac{\Delta G}{kT}\right) = \frac{kT}{h} \exp\left(-\frac{\Delta G^\dagger - \alpha G}{kT}\right). \tag{2}$$

In Eq. (2), k and h are the Boltzmann and Planck's constant,
respectively, T is the absolute temperature, $\Delta G^\dagger$ is the free energy
of activation in the absence of applied stress, and G the crack
driving force is the function of the stress, the specimen geometry
and the crack size a. Under the simplest conditions Eq. (1) is
valid even by itself — it is fortunate that the regions of validity
are quite wide. Further studies have shown [9] how an appropriate
combination of the rate constants can describe rigorously and completely SCC behavior over the full range of stress, crack velocity
and temperature. Thus, the rate theory approach to thermally activated fracture studies takes into consideration the atomic bond
energy of the material in the environment assisted cracking, as well
as the temperature. The introduction of the temperature as a variable extends the scope of fracture mechanics from the usual context
of loading and specimen geometry

$$t_{life} = g \text{ (load, geometry)}$$

to include the temperature also in the function

$$t_{life} = f \text{ (load, geometry, temperature)}$$

defined, of course, for the specific material and environment.

We wish to emphasize that our purpose is to develop an understanding of the construction and use of life expectancy topographs for design and test engineering practices: the present paper introduces the underlying theory and the concepts of topography construction to the research engineer. To simplify the mathematical derivation of the lifetime expression, the crack driving force G is used here in place of the more familiar stress intensity factor K.

THE DERIVATION OF THE LIFE-EXPECTANCY RELATION

The general behavior of materials in SCC is illustrated schematically in Fig. 1. It has been shown in previous studies |10-12| that each of the three regions are individually controlled by different rate constants, and that the crack velocity for the full range can be described as

$$v \approx \lambda(I,II) \frac{1}{|k(I)|^{-1} + |k(II)|^{-1}} + \lambda(III) \; k \; (III) \tag{3}$$

In Eq. (3), the rate constant k is the average number of crack tip movements per second, and the roman numerals assign the quantities to the appropriate region.

When the transition between the regions is wide as shown in Fig. 2a and the threshold zone affects the lifetime significantly (a frequent concern of the design engineer), a complex velocity relation of the form of Eq. (3) has to be considered in the determination of the critical time. (The complete velocity relation, which includes the threshold region as well, has been already reported |7,9|).

A sharp, well-defined transition between the regions, schematically represented in Fig. 2b, permits a simplified approach to the determination of the lifetime. In this case, Eq. (3) can be replaced by three separate velocity expressions, one for each of the three regions:

$$v_I = \lambda_I \frac{kT}{h} \exp(-\frac{\Delta G_I^\dagger}{kT}) \exp(\frac{\alpha_1 G}{kT})$$

$$= v_{0I} \exp(\frac{\alpha_I G}{kT}), \tag{4a}$$

$$v_{II} = v_{0II} \exp(\frac{\alpha_{II} G}{kT}), \tag{4b}$$

$$v_{III} = v_{0III} \exp(\frac{\alpha_{III} G}{kT}). \tag{4c}$$

134

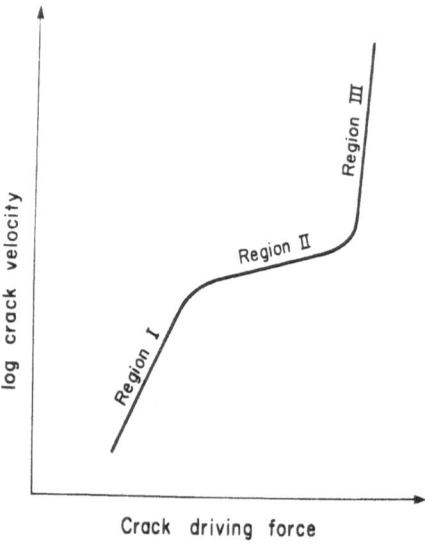

FIGURE I

Fig. 1. Schematic representation of the three regions of stress corrosion cracking.

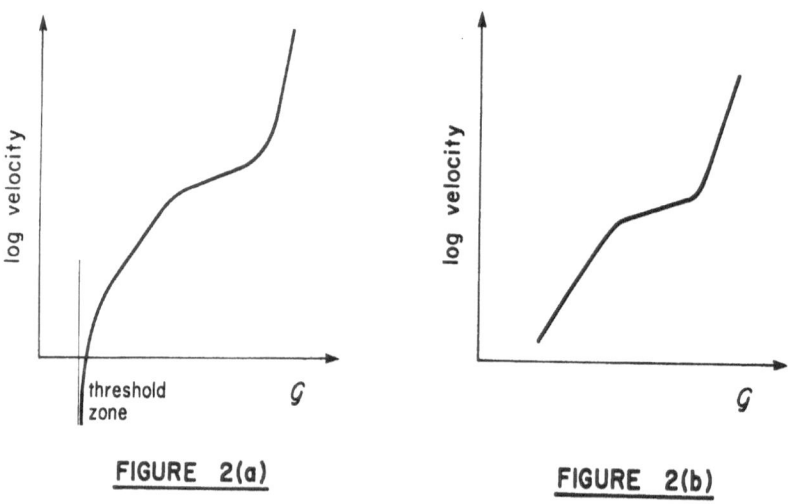

Fig. 2. Schematic representation of SCC diagrams (a) when the transitions between regions are wide, gradual and includes the threshold zone; (b) when the transitions are sharp, well-defined.

In Eqs. (4) the crack driving force G is

$$G = \sigma^2 \pi \, a/E, \tag{5}$$

where E is Young's modulus and, for constant stress σ

$$da = \frac{E}{\sigma^2 \pi} \, dG. \tag{6}$$

The time dt needed for the crack to grow by da is

$$dt = \frac{da}{v} . \tag{7}$$

Therefore, the time over which the crack grows from the initial size a_i to the critical size a_c is

$$\int_{t_i}^{t_c} dt = \int_{a_i}^{a_c} \frac{da}{v} = \int_{a_i}^{a_1} \frac{da}{v_I} + \int_{a_1}^{a_2} \frac{da}{v_{II}} + \int_{a_2}^{a_c} \frac{da}{v_{III}} . \tag{8}$$

Substituting Eq. (6) into Eq. (8) results in

$$\int_{t_i}^{t_c} dt = \frac{E}{\sigma^2 \pi} \int_{G_i}^{G_c} \frac{dG}{v} = \frac{E}{\sigma^2 \pi} \left(\int_{G_i}^{G_1} \frac{dG}{v_I} + \int_{G_1}^{G_2} \frac{dG}{v_{II}} + \int_{G_2}^{G_c} \frac{dG}{v_{III}} \right). \tag{9}$$

For Region I only, the limiting conditions are $t_i=0$, t_1, G_i, G_1, and the integration gives the lifetime as

$$t_1 - t_i = - \frac{Eh}{\sigma^2 \pi \, \alpha_I \, \lambda_I} \exp\left(\frac{\Delta G_I^{\dagger}}{kT} \right) \left| \exp\left(-\frac{\alpha_I G_1}{kT}\right) - \exp\left(-\frac{\alpha_I G_i}{kT}\right) \right|. \tag{10}$$

Carrying out the integration for Regions II and III in the same manner as for Region I with the corresponding constants of v_{II}, v_{III}, and between the integration limits of t_1, t_2, G_1, G_2; t_2, t_c, G_2, G_c, the combined, full lifetime expression is obtained:

$$t_c - t_i = (t_1 - t_i) + (t_2 - t_1) + (t_c - t_2)$$

$$= - \frac{Eh}{\sigma^2 \pi} \left\{ \frac{\exp\left(\dfrac{\Delta G_I^{\dagger}}{kT} \right)}{\alpha_I \, \lambda_I} \left| \exp\left(- \frac{\alpha_I G_1}{kT} \right) - \exp\left(- \frac{\alpha_I G_i}{kT} \right) \right| \right.$$

$$+ \frac{\exp(\frac{\Delta G^{\dagger}_{II}}{kT})}{\alpha_{II} \lambda_{II}} \left| \exp(- \frac{\alpha_{II} G_2}{kT}) - \exp(- \frac{\alpha_{II} G_1}{kT}) \right|$$

$$+ \frac{\exp(\frac{\Delta G^{\dagger}_{III}}{kT})}{\alpha_{III} \lambda_{III}} \left| \exp(- \frac{\alpha_{III} G_c}{kT}) - \exp(- \frac{\alpha_{III} G_2}{kT}) \right| \}. \tag{11}$$

Based on the physically rigorous theory of thermally activated processes, the mathematical analysis of SCC leads to the lifetime expression Eq. (11), from which the geometrical form can be constructed. Compared to the mathematical description of lifetime predictions, the great advantage of the three-dimensional representation is in its easy readability for design, test, and reliability engineering purposes |13|. The detailed step-by-step procedure for the conversion of the mathematical relation into the three-dimensional diagram of the lifetime properties of SCC is given here. To facilitate the understanding of the construction, the three regions will be bounded by flat planes in the G_i - t_c - T coordinate system of the present study: the expected shape of the surface for Region I is indicated also.

THE EXPECTED SHAPE OF THE TOPOGRAPH

1. The trace on the G_i - T plane. From Eq. (10)

$$t_1 \propto - \exp(\frac{\Delta G^{\dagger}_I - \alpha_I G_1}{kT}) + \exp(\frac{\Delta G^{\dagger}_I - \alpha_I G_i}{kT}),$$

or

$$t_1 \propto - \frac{1}{v_1} + \frac{1}{v_i}.$$

For sufficiently small initial crack driving force G_i, it is usually found from the log v vs G diagrams that v_1 is greater than v_i by orders of magnitude. In this case $G_1 \gg G_i$, and consequently

$$\exp(\frac{\Delta G^{\dagger}_I - \alpha_I G_1}{kT}) \ll \exp(\frac{\Delta G^{\dagger}_I - \alpha_I G_i}{kT})$$

to the extent that $\frac{1}{v_1}$ is negligible and Eq. (10) becomes

$$t_1 \simeq \frac{1}{C_I} \exp(\frac{\Delta G^{\dagger}_I - \alpha_I G_i}{kT}). \tag{12}$$

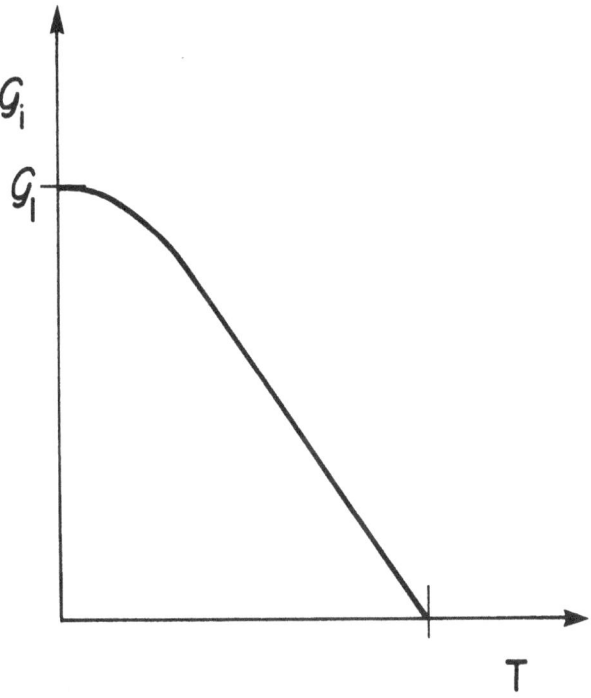

FIGURE 3(a)

Fig. 3. Developing from Eq. (10) the expected shape of the topograph for Region I. (a) and (b) the trace on the

G_i - T plane, with $T = \dfrac{\Delta G_I^{\ddagger} - \alpha_I G_i}{k \ \ell n t_1 C_I}$. Note that the time

and temperature coordinates do not have zero values at the origin of the coordinate system. (c) the trace on the G_i - t_1 plane. (d) the trace on the t_1 - T plane, with negative and decreasing slope for decreasing temperature. (e) the expected surface, assembled for Region I of stress corrosion experimental results.

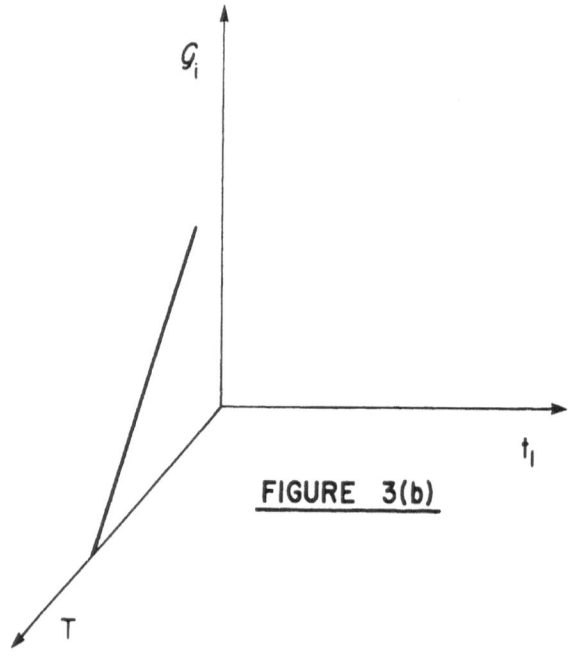

FIGURE 3(b)

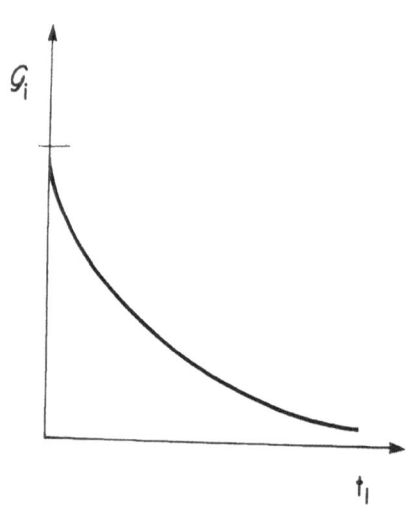

FIGURE 3(c)

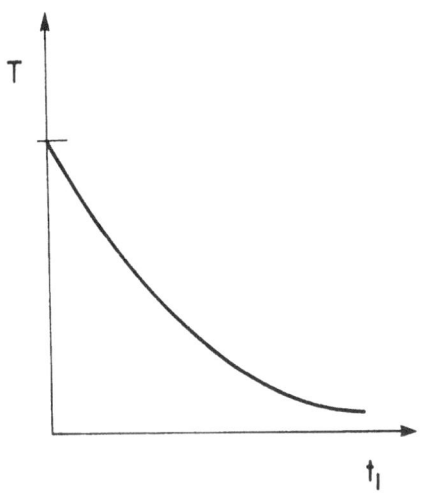

FIGURE 3(d)

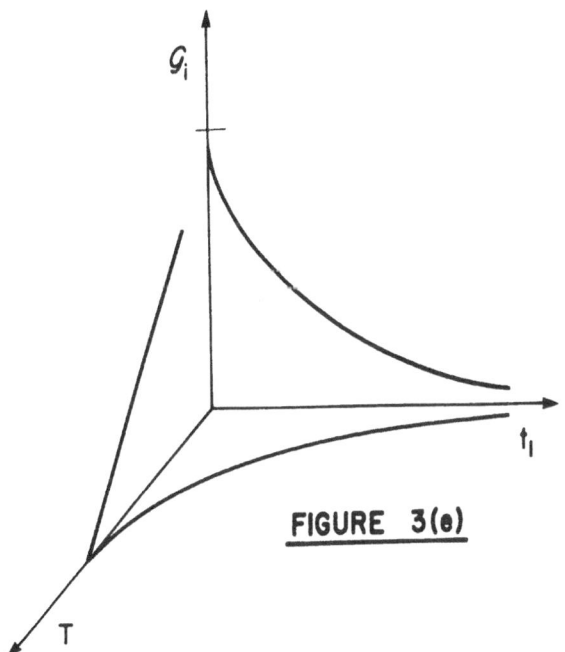

FIGURE 3(e)

Rearranging Eq. (12) results in the simple form

$$G_i = C_I' - C_I'' T,$$

showing that the initial crack driving force is a linear function of the temperature with a negative slope when $G_1 \gg G_i$. When G_i is near the value of G_1 the trace of the surface will be probably of the type as shown in Fig. 3a. The exact shape of the trace, however, is subject to the condition that G_1 is independent of the temperature. Unless this independence is proven or experimentally observed, the theoretical topograph should have a straight edge in the G_i - T plane (Fig. 3b).

2. The trace on the G_i - t plane. As before, from Eq. (10)

$$t_1 \propto - \exp(\frac{\alpha_I}{kT} G_1) + \exp(- \frac{\alpha_I}{kT} G_i).$$

Since at constant temperature G_1 has a single value only, it follows immediately that the lifetime changes as

$$y = \exp(- b x),$$

and the trace of the surface will be as shown in Fig. 3c.

3. The trace on the T - t_1 plane. To find the shape of the trace, it is enough to investigate how the slope changes with the temperature. If Eq. (10) is expressed as

$$t_1 = - C \exp(\frac{\Delta G_I^{\dagger} - \alpha_I G_1}{kT}) + C \exp(\frac{\Delta G_I^{\dagger} - \alpha_I G_i}{kT}),$$

then,

$$\frac{dt_1}{dT} = \frac{dt_1}{d1/T} \frac{d1/T}{dT} = - C \frac{\Delta G_1}{kT^2} \exp(- \frac{\Delta G_1}{kT}) + C \frac{\Delta G_i}{kT^2} \exp(- \frac{\Delta G_i}{kT}).$$

Again, as for the G_i - T plane, consider the low range of G_i, so that $\exp(- \frac{\Delta G_1}{kT})$ is negligible, and

$$\frac{dt_1}{dT} \approx C \frac{\Delta G_i}{kT^2} \exp(- \frac{\Delta G_i}{kT}).$$

The behavior of this function is such that as the temperature decreases the slope decreases also and is always negative, as shown in Fig. 3d. Figure 3e is the expected shape of the surface, in the G_i - t_1 - T coordinate system, for Region I only.

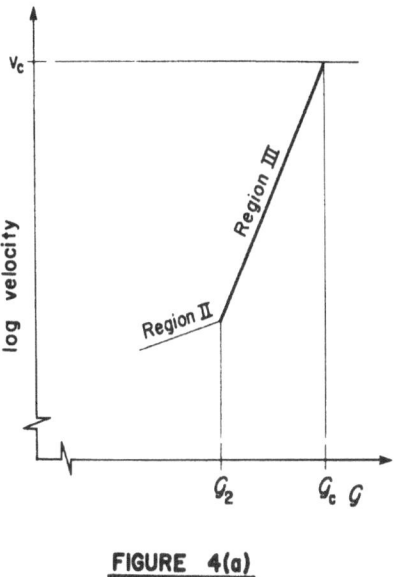

FIGURE 4(a)

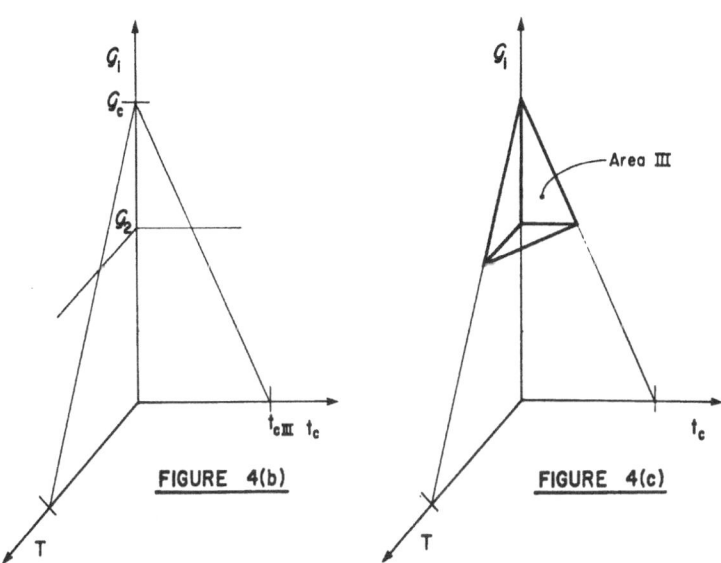

FIGURE 4(b)   **FIGURE 4(c)**

Fig. 4. (a), (b), (c). The step-by-step construction of Area III for Region III of stress corrosion.

142

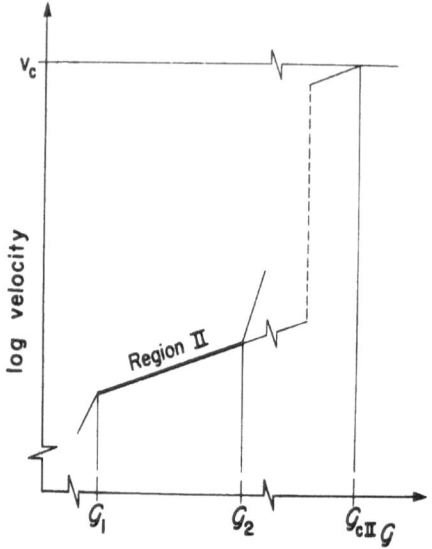

FIGURE 5(a)

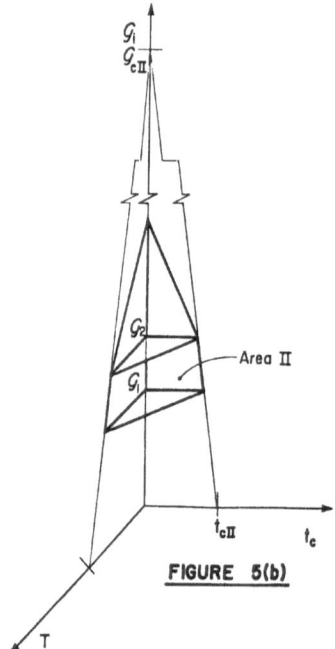

FIGURE 5(b)

Fig. 5. (a), (b). The construction of Area II.

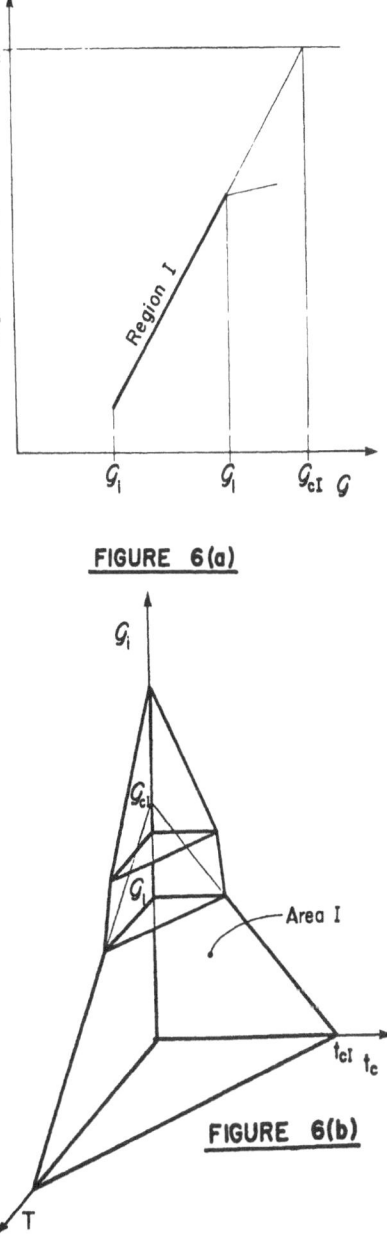

FIGURE 6(a)

FIGURE 6(b)

Fig. 6. (a), (b). The construction of Area I and the assembled three-dimensional schematic representation of the life expectancy calculation, Eq. (11).

THE CONSTRUCTION OF THE TOPOGRAPH

The complete topograph will now be assembled. The construction should start with Region III as shown in Fig. 4a. The crack driving force values G_c and G_2, labelled in accordance with the limiting values of Eq. (11), give the apex and the base of Area III. The experimental temperature and the critical time when failure occurs, that is, the life expectancy value determines tie-lines G_c - T and G_c - t_{cIII}, Fig. 4b. The T - t_{cIII} tie-line is at the constant crack driving force level G_2, and the heavy outline in Fig. 4c shows the area of the topograph that is valid for Region III.

To construct Area II, Region II has to be extrapolated to the measured critical velocity value, as shown in Fig. 5a, to obtain the "theoretical critical crack driving force" G_{cII}, giving the apex of the area on the G_i axis. The traces on the G_i - T and G_i - t_c planes start at that point and join Area III at its base, at G_2. The base of Area II is at G_1 — the validity of Area II is limited by this constant crack driving force level, as shown in Fig. 5b.

The construction of Area I also has to start with the extrapolation of Region I to v_c, as shown in Fig. 6a. As for Area II, from the "theoretical crack driving force" G_{cI} so obtained the apex is found, giving the starting point for the traces on the G_i - t_c and G_i - T planes. At G_1 the traces join — and thus cut — Area II, and the Area that is valid for Region I can be outlined. The fully assembled topograph is shown in Fig. 6b.

CONCLUSIONS

Because SCC is a thermally activated process, the studies of environmentally assisted crack propagation must be based on the theory of thermally activated rate processes. As it has been shown, the experimental results can be rigorously described by the combination of rate theory and fracture mechanics. The rate theory approach extends the scope of fracture mechanics through the consideration of the temperature, and leads to crack velocity relations that can be integrated directly to obtain the life expectancy of the material. This, in turn, lends itself to three-dimensional representation in the crack driving force — temperature — lifetime coordinate system, giving vital information in a form that is then readily available for engineering purposes. The topography diagrams of different alloys that were subjected to the same environment can be superposed, and the appropriate material can be selected immediately for the planned G_i - T - t_c combination.

In the foregoing, it was considered that the component was designed to operate at constant nominal stress, temperature, and environment. Failure will occur when the crack size reaches a critical value a_c, corresponding to the critical value of the crack driving force that is a characteristic of the material — environment system (including the temperature).

It was also considered in the development of Eq. (11) that the initial crack sizes are uniform. However, in all engineering materials, a spectrum of initial crack sizes exists. Furthermore, when the applied stress is not constant, as it is frequently the case, a distribution of service life exists. These aspects are now under active study.

ACKNOWLEDGEMENTS

The generous hospitality extended to both authors by the members of the Laboratoire de Mécanique des Solides, Ecole Polytechnique, Palaiseau, France, and the financial assistance, provided by the Natural Sciences and Engineering Research Council of Canada (to A.S. Krausz), are gratefully acknowledged.

REFERENCES

1. A.S. Krausz, Journal of Applied Physics, 49(1978), 3774.
2. S.W. Wiederhorn, Journal of the American Ceramic Society, 50(1967), 407.
3. B.R. Lawn, Journal of Material Science, 10(1975), 469.
4. W.B. Hillig and R.J. Charles in "High Strength Materials" (Ed. V.F. Zackey), Wiley, New York (1965).
5. H.M. Cekirge, W.R. Tyson and A.S. Krausz, Journal of American Ceramic Society, 58(1976), 265.
6. W.R. Tyson, H.M. Cekirge and A.S. Krausz, Journal of Material Science, 11(1976), 780.
7. A.S. Krausz, International Journal of Fracture, 14(1978), 5.
8. A.S. Krausz and H. Eyring "Deformation Kinetics", Wiley-Interscience, New York (1975).
9. A.S. Krausz, Engineering Fracture Mechanics, 11(1979), 33.
10. S.D. Brown, Proceedings of the 6th Canadian Congress of Applied Mechanics, Vancouver, B.C. (1977).
11. A.S. Krausz, and K. Krausz, Engineering Fracture Mechanics, 13(1980), 751.
12. A.S. Krausz, J. Mshana and K. Krausz, Engineering Fracture Mechanics, 13(1980), 759.
13. D. Broek "Elementary Engineering Fracture Mechanics", Sijthoff & Noordhoff, Alphen aan den Rijn — The Netherlands (1978).

EFFECTS OF HYDROGEN AND HEAT TREATMENT CONDITIONS ON FRACTURE
TOUGHNESS (J-INTEGRAL) OF QUENCHED AND TEMPERED SAE-4130 STEEL

P. NGUYEN-DUY and S. LALONDE

Institut de recherche d'Hydro-Québec (IREQ)
Varennes, Québec, Canada, J0L 2P0

ABSTRACT

Elastic partial unloading J_R curve method for the determination of J-Integral critical value, J_{IC}, was used to study the
effect of hydrogen and heat treatment conditions on fracture
toughness of SAE-4130 steel. Three tempering temperatures were
selected (400°C, 500°C and 600°C). J_{IC} values increased drastically from 400°C (J_{IC} = 0,012 MJ-m^{-2}) to 500°C (J_{IC} = 0,134 MJ-
m^{-2}) but remained nearly constant as tempering temperature increased from 500°C to 600°C. Specimens tempered at 500°C and
600°C were submitted to electrolytic hydrogen charging during the
J_{IC} testing. The susceptibility to hydrogen embrittlement of
SAE-4130 for these tempering conditions could be clearly observed
as J_{IC} decreased for increasing charging times. A difference was
noted in the nature of the fracture surface for embrittled and
unembrittled specimens.

1.   INTRODUCTION

The absorption of hydrogen by steel is relatively easy and
depends on different factors which are intimately related to chemical composition, mechanical properties and characteristics of
the material. There are presently three different theories which
explain hydrogen embrittlement: the internal pressure theory, the
decohesion theory and the adsorption theory[1-2].

The internal pressure theory relates the embrittling effect
of hydrogen to the occurence and growth of internal cavities.
These cavities are grain boundaries, matrix-inclusion interfaces

or porosities where hydrogen can precipitate and create on increase in pressure that will lead to the propagation of a crack.
External stresses would also have an effect on the crack pattern[3].
In the decohesion theory, hydrogen atoms in solid solution inter-
fere with the metallic bond of a transition metal. This disrup-
tion would come from an interaction in the 3d electron band, resulting in higher repulsive forces or from an interaction in the
4s electron bond, resulting in a more anisotropic bond. In the
first case, there is a loss of cohesion due to greater equilibrium
distances between atoms[4]. In the second case, a chemical bond
would be formed between hydrogen and iron resulting in a more
directional bond between metallic atoms[5]. The metal would then
have a brittle behaviour. The adsorption theory (or surface
energy theory) is based on the existence of microcrack for embrittlement to occur; in this theory, hydrogen must be immediately
adsorbed an internal surface, at the tip of a microcrack thus reducing surface energy.

Chemical composition and microstructure have a great effect
on hydrogen embrittlement[6]. The fracture toughness of embrittled
materials decreases and the nature of these materials changes
drastically.

Many fracture mechanic criteria are available to evaluate
fracture toughness of materials, J-Integral is one of them. J-
Integral can be calculated from results obtained from multi-
specimens method or single specimen method. The elastic partial
unloading method, is a single specimen method which allows the
calculation of J-Integral critical value: J_{IC}. In the present
study, J_{IC} is determined for each tempering temperature wether the
specimen is embrittled by hydrogen or not. A computer program had
previously been written to control the testing machine and to monitor the value of J during a test. This value can be calculated
using the following equation developed for pure bending[7].

$$J_{i+1} = \left[J_i + \left(\frac{2A_{i,i+1}}{b_i\, B} \right) \right] \left[1 - \left(\left(\frac{1+0,76(\frac{b_i}{W})}{b_i} \right)(a_{i+1}-a_1) \right) \right] \tag{1}$$

In equation 1, the crack extension $(a_{i+1}-a_i)$ is taken into
account for the calculation of J. For pure bending condition, a
compliance equation exists enabling the evaluation of $(a_{i+1}-a_i)$
while determining the compliance of the specimen $(\Delta COD/\Delta P)$ for
each unloading. For each test, data obtained are load (P) - load
line displacement (δ), load (P) - crack mouth oepning displacement
(COD). A regression analysis performed during the test on the
variation of P-COD leads to the evaluation of $\Delta COD/\Delta P$ for each
unloading and then to crack extension $\Delta a_{i+1,i} = a_{i+1}-a_i$. The
variation of P-δ gives the possibility to calculate the energy

associated with each unloading which is the area under the curve:

$$A_{i+1,i} = \int_{\delta_i}^{\delta_{i+1}} P d\delta \qquad (2)$$

The two sets of data will be used to calculate J-Integral by equation (1). It is also important to mention that the regression analysis was performed on 60 points, centrally located for each unloading on the P-COD curve. Twenty points at the beginning and at the end of each unloading were ignored in the data set. Unloadings were realized for 15% of the actual load.

2. EXPERIMENTAL PROCEDURES

2.1 Material and Specimens

The material used in this study was quenched and tempered SAE-4130 steel. Chemical analysis was performed and results are reported in Table 1. It can be seen from those results that heat treating conditions did not significantly alter the steel composition.

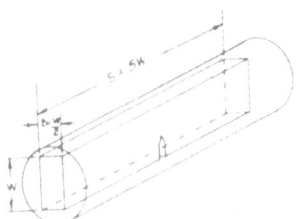

Fig. 1: Location of four-points bend specimens

Specimens were machined from 4,45 cm diameter bars previously heat treated (quenched and tempered at 590°C). Fig. 1 shows how specimens were cut from a bar. Dimensions are as specified in Fig. 2 where:

$S_1 = 2 \; S_2 = 4 \; W = 8 \; B = 128$ mm

the initial mechanical notch depth is 10 mm. Six tensile specimens were machined from the same material for the determination of conventional mechanical properties (tensile properties) at different tempering temperatures. Microstructures of steel for the three tempering temperatures are shown in Fig. 3.

150

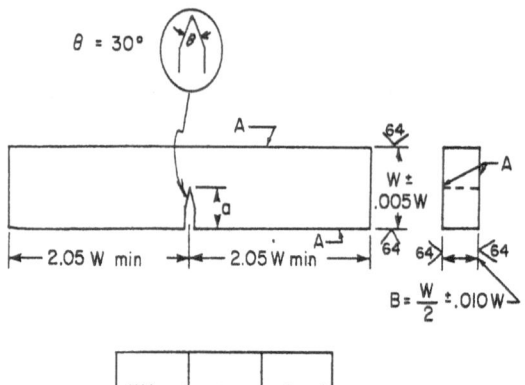

Fig. 2: Geometry and dimensions of specimens.

Table 1: Chemical composition of SAE-4130 steel.

STEELS (4130)	C	Mn	P	S	Si	Cu	Ni	Cr	Mo	V	Al	N ppm
NOM.	.28 .33	40 60	035 max	.040 max	15 30			.80 1.10	15 25			
REC.	.34	64	014	024	22	12	.06	1 05	16	012	022	
400	36	61	030	.037	24	11	.07	.98	178		017	100
500	32	60	.011	029	20	16	.06	88	168		022	93
600	35	61	.028	037	23	.11	.07	.98	180		018	102

2.2 Experiments

For these experiments, we chose to vary two parameters:

— the tempering temperature, T_t

— the duration of electrochemical charging, t_H

— all other parameters, such as austenitization time and temperature, tempering time or solution composition and temperature remained unchanged.

Heat Treating: After austenitization at 870°C for 45 minutes

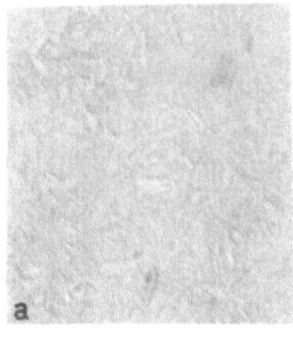

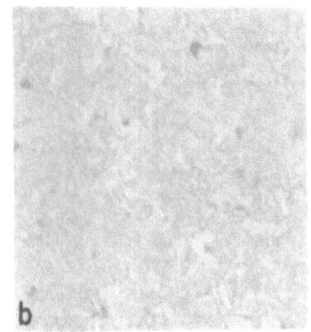

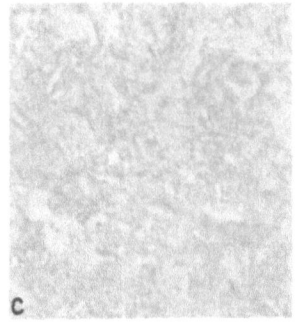

a) $400^{\circ}C$ 1000 X

b) $500^{\circ}C$ 1000 X

c) $600^{\circ}C$ 1000 X

Fig. 3: Microstructures of SAE 4130 steel for the three dif-
ferent tempering temperatures ($400^{\circ}C$, $500^{\circ}C$ and $600^{\circ}C$).

in a liquid salt bath, specimens were oil quenched (at 25ºC) wash-
ed and tempered without delay, at three different temperatures
(400ºC, 500ºC and 600ºC) for 60 minutes. Finally, specimens were
allowed to air cool. Untreated specimens (in the "as received" condition) were also used in order to determine J_{1c}. This constitute
the fourth and last condition (Fig. 4).

Surface Treatment: All specimens were then polished to 3 µm
to remove the oxide scale which influences greatly the adsorption
of hydrogen by poisoning recombination reactions[8]. This polishing
creates a plastic deformation in a thin surface layer which acts
as a trap for hydrogen and slows down diffusion[9], for this reason,
the plastically deformed layer must be chemically removed. Etching
reagent used for this purpose had previously been described by
Hirth[10].

Precracking: The partial unloading method uses precracked
specimens. The precracking was done on a rigid four-point-bend
fixture allowing pure bending. The small span S_2 and the large

Fig. 4: Microstructure of as received
quenched and tempered SAE-4130
steel, 450 X.

span S_1 are respectively 64 mm and 128 mm. All specimens were
fatigue precracked in the range of

$$0,50 \leqslant a_{0/W} \leqslant 0,65$$

Experimental Set-Up: The experimental set-up is shown in
Fig. 5, an inverted four-point bend set-up is used, enabling the
utilization of a normal COD gage. Two platinum electrodes were
placed in the cell, facing the lateral surfaces of the specimen
and at the level of the fatigue precrack. A 0,1 N sulfuric acid
solution added with a few drops of a 1 g/l thiourea solution was
used as an embrittling agent. This solution is freshly made for
each test. Specimens were cathodically polarized with a current
density of 24 mA/cm².

Fig. 5: Experimental set-up.

3. RESULTS

3.1 Effects of Tempering Conditions

Results obtained on different specimens for different tempering temperatures are shown in Table 2 and Figure 6.

The elastic partial unloading method used to determine the J-Integral on specimen tempered at different temperatures clearly shows that at a tempering temperature of 400°C, crack onset and crack propagation is very brittle, specimens do not accomodate any plastic deformation. This can be easily observed on fracture surfaces (Figs. 7,8) and on the load vs displacement curve (Fig. 9). The fracture surfaces exhibit a brittle behaviour, with inter-granular crack propagation mode in the case of fatigue precrack

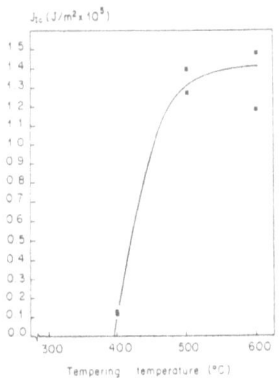

Fig. 6: Variation of J_{Ic} as a function of tempering temperature for SAE-4130 steel specimens.

and transgranular and intergranular crack propagation mode for slow bend propagation. Those specimens (T_t = 400°C) give a low value of J_{Ic} which are 0,0123 and 0,0113 MJ-m^{-2} as the tempering temperature increases from 400°C to 500°C, the fracture mode seems to undergo transition from brittle to ductile behaviour. Similarly, J_{Ic} values increase drastically and the lower limit of temperature is nearly 500°C. Between 500°C and 600°C, the fracture toughness of SAE-4130 steel varies slightly.

The fracture surfaces of specimens tempered at 500°C and 600°C are ductile as shown in figures 10 and 11, contrasting with those of specimens tempered at 400°C. The critical values of J obtained are: 0,127 and 0,138 MJ-m^{-2} when tempering temperature is 500°C and 0,149 and 0,118 MJ-m^{-2} when tempering temperature is 600°C. These values are comparable to the value previously obtained for specimens cut from identical bars and tested in the "as received" condition (J_{Ic} = 0,1414 MJ-m^{-2})[11].

154

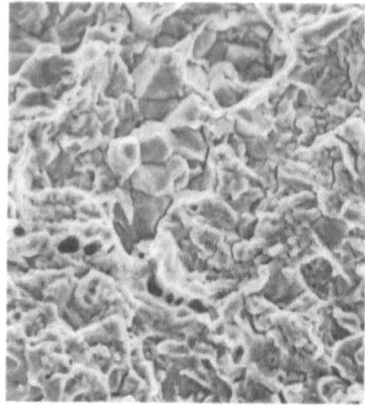

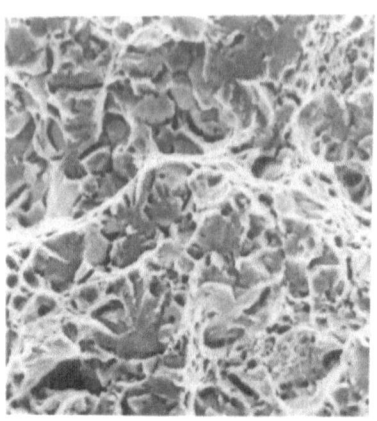

Fig. 7: Fracture surface of specimen tempered at 400°C, fatigue zone (1000 X).

Fig. 8: Fracture surface of specimen tempered at 400°C, slow bend crack propagation (1000 X).

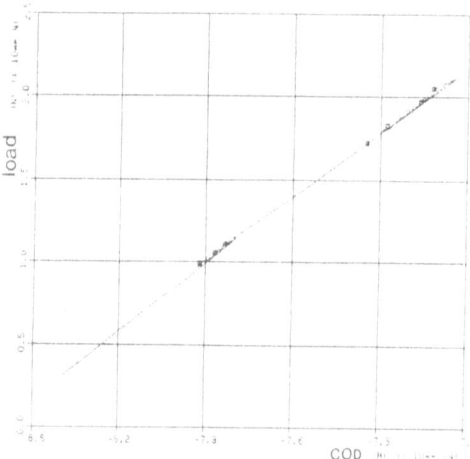

Fig. 9: Load-crack opening displacement curve for specimen tempered at 400°C.

3.2 Effects of Hydrogen Charging

Four specimens tempered at 500°C and six specimens tempered at 600°C were electrochemically loaded with hydrogen during the elastic partial unloading test to determine the J-Integral. The variation of J_{IC} as a function of the electrochemical charging time, t_H, is illustrated in figure 12. Results appear in Table 2.

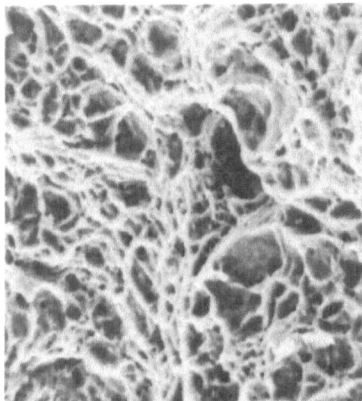 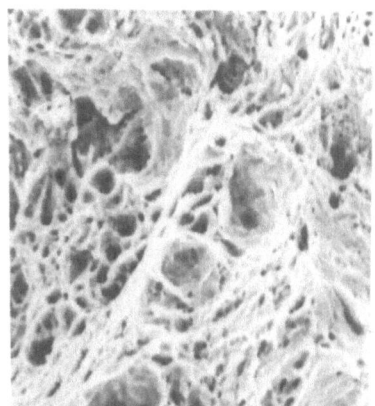

Fig.10: Fracture surface of specimen tempered at 500°C (1500 X).

Fig.11: Fracture surface of specimen tempered at 600°C (1500 X).

Table 2: J_{IC} values of specimens tempered at different temperature for different hydrogen charging duration.

TEMPERING TEMPERATURES (°C)	HYDROGEN CHARGING TIMES (min)	J_{IC} (MJ-m⁻²)
400	0 00	0 0123
400	0 00	0 0113
500	0 00	0 1273
500	0 00	0 1386
500	25 87	0 1050
500	40 79	0 0794
500	59 91	0 0479
500	84 93	0 1318
590	0 00	0 1414
600	0 00	0 1480
600	0 00	0 1183
600	24 34	0 1400
600	51 78	0 0990
600	55 31	0 1550
600	72 06	0 0427
600	94 19	0 0356
600	133 00	0 0753

Although there is a large discrepancy among the results, especially for the specimens tempered at 600°C, they show a general tendency to decrease as the charging time increases specimens made from SAE-4130 quenched and tempered steel (500°C and 600°C), are thus susceptible to hydrogen embrittlement. J_{IC} values decreased from 0,133 MJ-m⁻² to a value between 0,0425 MJ-m⁻² and 0,0794 MJ-m⁻² for a tempering temperature of 500°C and different charging times. This decreasing tendency is less significant in the case of specimens tempered at T_t = 600°C. J_{IC} varies from 0,133 MJ-m⁻², for unembrittled specimens, to 0,075 MJ-m⁻² for specimens charged with hydrogen during 133 min. Significantly lower values are nevertheless observed for shorter charging times.

156

The embrittling effects of hydrogen were very easily noticed on the fracture surface. As shown in figures 13 and 14, respectively for specimens tempered at 500°C and 600°C, the fracture surface in these cases are different from those obtained from unembrittled specimens, especially in the case of specimens tempered at 500°C, where the fracture appearance is brittle. Many secondary cracks were observed as can be seen in figure 15. For specimens with low J_{Ic}, a tendency to delaminate is strong and the fracture surface has a fibrous appearance. Small cracks tend to propagate perpendicularly to the principal surface fracture along inclusions line. Examples are shown in figures 16 and 17.

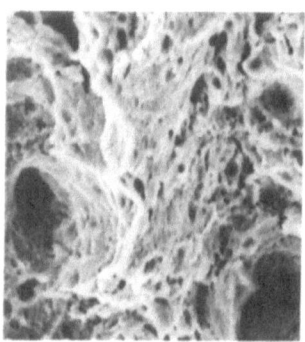

Fig.13: Fracture surface of an embrittled specimen, tempered at 500°C (2000 X).

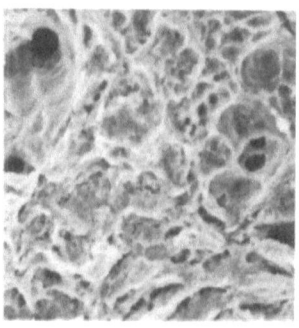

Fig.14: Fracture surface of an embrittled specimen, tempered at 600°C (1500 X).

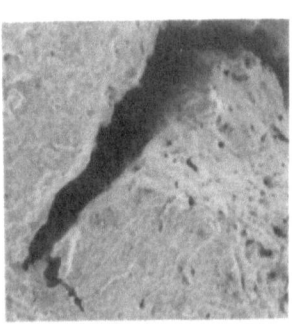

Fig.15: Secondary cracks in a specimen tempered at 500°C (100 X).

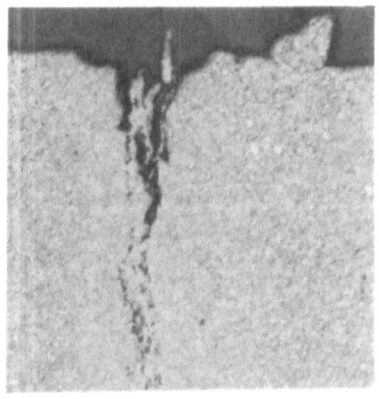

Fig.16: Delamination taking place along inclusions line (50 X).

Fig.17: Fracture surface for the specimen shown in Fig. 16. Tempered at 500°C (100 X).

4. CONCLUSIONS

The elastic partial unloading J-R curve method was successfully used to detect the effects of tempering temperature and hydrogen on fracture toughness of SAE-4130 steel.

A brittle-ductile transition was observed in the behaviour of J_{Ic} with increasing tempering temperature. For SAE-4130 steel, this transition is in the range of 450°C-500°C where 500°C is the beginning of the uppershelf.

SAE-4130 steel tempered at 500°C and 600°C shows clearly a susceptibility to hydrogen embrittlement. Fracture toughness quantified by J_{Ic} decreased with the presence of hydrogen.

In the presence of hydrogen, many secondary cracks and delimination cracks can be observed. The delamination cracks take place perpendicularly to the principal fracture surface, along inclusions lines. It is suggested that embrittling mechanisms involve a cooperation between hydrogen and hard inclusions particles. It suggests that the material cleanliness is an important factor that influences the susceptibility to hydrogen embrittlement.

ACKNOWLEDGEMENTS

The authors are very grateful to professor I. Dickson of Ecole Polytechnique for many helpful discussions and recommendations. The authors would also like to thank R. Bégin, R. Dufour, C. Lanouette and J.G. Millette for their technical assistance and paper preparation. Authors are indebted to J. de Maisonneuve for typing the manuscript.

158

REFERENCES

1. Hirth, J. P. Effects of Hydrogen on the Properties of Iron
   and Steel. Metallurgical Transactions a. 11A (1980) 861-890.

2. Pressouyre, G. M., Dollet, J. et Vieillard-Baron, B. Evolution
   des Connaissances concernant la Fragilisation par l'Hydrogène
   des Aciers. Mémoires et Etudes Scientifiques, Revue de la Métallurgie. 79(4) (1982) 161-176.

3. Iino, M. The Extension of Hydrogen Blister Crack Array in Line Pipe Steels. Metallurgical Transactions A. 9A(1978) 1581-
   1590.

4. Oriani, R. A. A Mechanistic Theory of Hydrogen Embrittlement
   of Steels. Berichte der Bunsen-Gellschaft fur Physikalische
   Chemie. 76(1972) 301-310.

5. Losch, W. Hydrogen Embrittlement: a new Model for the Mechanism of Reduction of Metallic Cohesion. Scripta Metallurgica.
   13(8)(1973) 661-664.

6. Lalonde, S. Unpublished Research, Ecole Polytechnique de Montréal, 1984.

7. Ernst, H. A., Paris, P. C. and Landes, J. D. Estimations on
   J-Integral and Tearing Modulus from a Single Test Record.
   Fracture Mechanics: Thirteenth Conference, ASTM STP743, Richar.l
   Roberts Ed., American Society for Testing and Materials.
   (1981) 476-502.

8. Heubaum, F. H. and Berkowitz B. J. The Effect of Surface
   Oxides on the Hydrogen Permeation Through Steels. Scripta
   Metallurgica. 16(6) (1982) 659-662.

9. Hill, M. L. and Johnson, E. W. Hydrogen in Cold Work Iron-
   Carbon Alloys and the Mechanisms of Hydrogen Embrittlement.
   Transaction of the Metallurgical Society of AIME. 215 (1959)
   717-725.

10. Onyewuenyi, O. A. and Hirth, J. P. The Effect of Hydrogen on
    Microhardness of Spheroidized AISI 1090 Steel. Scripta Metal-
    lurgica. 15(1981) 113-118.

11. Nguyen-Duy, P. and Bayard, S. Fracture Toughness of 4130
    Quenched and Tempered Steel. Journal of Engineering Materials
    and Technology. 103(1981) 55-61.

INFLUENCE OF INTERFACIAL ADHESION ON FATIGUE OF COMACO COMPOSITES

P.W.K. Lam and M.R. Piggott

Department of Chemical Engineering & Applied Chemistry
University of Toronto, Toronto, Ontario M5S 1A4

ABSTRACT

Fatigue and Izod tests have been carried out on aligned carbon fibre composites in which the cure shrinkage of the matrix was controlled. Both fatigue resistance and toughness correlated with shrinkage pressure. There was negligible loss of Young's modulus during fatigue, up to 80% of the fatigue life, while the flexural modulus declined continuously during fatigue. A hole, drilled in the specimen, increased in diameter at right angles to the stress during the fatigue. Both rate of flexural modulus decline and rate of increase in hole size correlated with shrinkage pressure. These results are interpreted on the basis of the improvement of the adhesion between fibres and matrix that is brought about by the reduction in resin shrinkage.

INTRODUCTION

Fibre reinforced plastics are being widely applied, and the high performance types are having a large effect on the aerospace industry. This is because their high strength and modulus to density ratios result in more efficient structures. In addition, the material can be designed to have all that is required in strength and stiffness, and no more. This design potentiality brings with it an even greater structural efficiency, and is leading to the virtual complete replacement of aluminum in aircraft structures. The all-carbon reinforced plastic structure is already a reality for small aircraft, and large passenger aircraft will quickly follow this trend.

In one respect, however, carbon reinforced plastics are in
need of improvement. They lack ductility, and the associated brittleness has two aspects: 1) delamination, with consequent loss of
flexural and compression strength due to impact by blunt objects,
and 2) low fracture toughness, as emplified by Izod and Charpy,
or K_{IC} tests. The first problem is being tackled by the use of
tougher resins, but the second, although studied for much longer,
has only recently seen much progress.

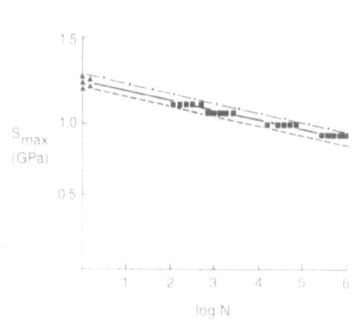

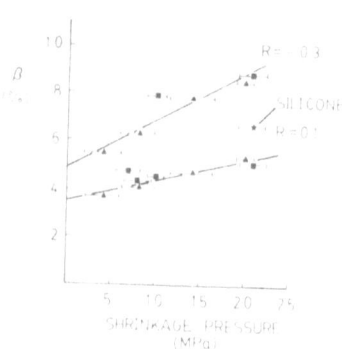

Fig. 1. S-N curve for pultrusion.   Fig. 2. Slope of S-N curve.
         The dashed and dotted lines β, vs matrix shrin-
         indicate 0.02 and 0.98   age pressure.
         failure probability
         respectively.

The major contributor to fracture toughness of fibre composites
(Kevlar fibre composites excepted) arises directly neither from
the matrix, nor the fibres. In fact, a ductile and tough aluminum,
when reinforced with carbon fibres, produces a very brittle composite. This is because the fibres inhibit the plastic flow in the
aluminum at the crack tip (1). Toughness arises, instead, from
interactions between fibres and matrix. The most important of
these is fibre pull out (2).

The work of fibre pull out is governed by the interfacial
shear stress, τ_i, and, when long fibres are used, is greatest when
τ_i is very small (3). Unfortunately, when τ_i is small the shear
strength (4) and compression strength (5) of the composites are
low. For many years, this presented an impasse.

Recently, however, it was shown theoretically (6) that control of the polymer cure shrinkage could increase composite toughness. This was verified by experiment (7) where it was also shown

that the improvement was obtained without loss of shear strength. The method was patented (8), and this type of composite bears the tradename COMACO Composite.

This paper describes some work on the fatigue properties of these composites.

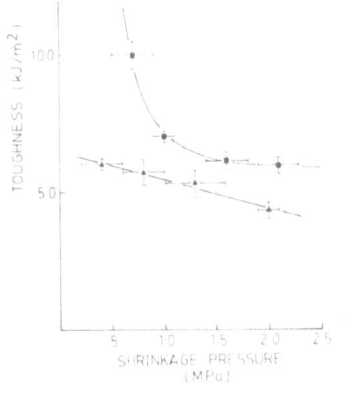
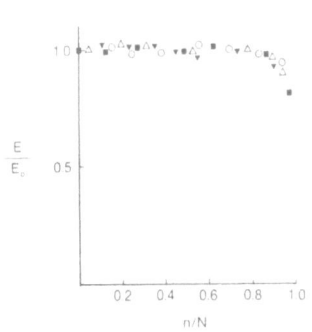

Fig. 3. Izod toughness vs matrix
shrinkage pressure. ▲;
additive A: ■ ; additive B.

Fig. 4. Loss of Young's
modulus during
fatigue. R = 0.1

EXPERIMENTAL METHOD

The fibre composites were made with Hercules AS1 carbon fibres, and a number of different epoxy resins. Samples were pultruded, and cured inside the molds. They were end tabbed and tested in a servo hydraulic machine, at a frequency of 10 Hz, except for those that had fatigue lives less than 1000 cycles. These were tested at 2Hz to avoid specimen heating. Tests were carried out in tension (R = 0.1) and tension-compression (R = -0.3). Static tests were also carried out similarly, and toughness was measured using the Izod test.

The epoxy resins were modified, using laboratory synthesized additives which copolymerized with the epoxy resin, and in doing so expanded. By using different amounts of these additives, the cure shrinkages of the resins were reduced in a controlled manner.

162

In some of the experiments the fibres were coated with silicone oil by immersing the strands in a 5% solution of the oil,
and drying before use.

EXPERIMENTAL RESULTS

The S-N curves were linear, fig. 1, and were characterized
by their slopes, β% per decade. β was strongly dependent on the
matrix cure shrinkage pressure, fig. 2, as estimated by using
strain gauges. The Izod toughness was also dependent on shrinkage
pressure, fig. 3.

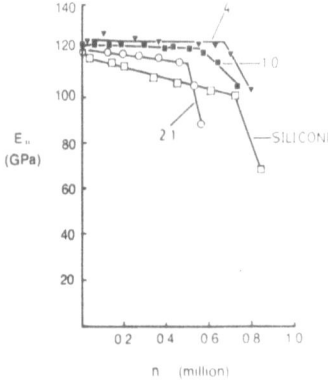
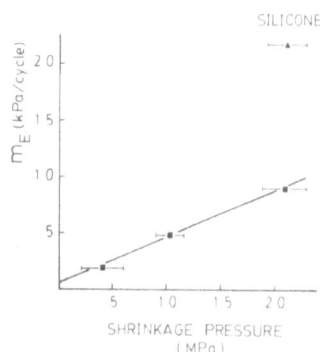

Fig. 5.   Loss of flexural   Fig. 6.   Rate of loss of flexural
          modulus during   modulus vs slope of
          fatigue. R = 0.1.   S-N curve. R = 0.1

The Young's modulus of the composites did not decline significantly until almost the end of the fatigue life, fig. 4, whereas the flexural modulus declined quite noticeably, fig. 5. The
decline in flexural modulus correlated with shrinkage pressure,
fig. 6.

Specimens which had a 4.6 mm diameter hole cut in them were
also fatigue tested. It was observed that the hole size increased
only a little in the direction of the applied stress (and fibres),
but increased very significantly at right angles to the stress,
fig. 7, for R = 0.1. With R = -0.3 similar results were obtained,
but the changes in hole dimensions were greater, fig. 8. In
both cases the increase at right angles to the stress was greater
for the composites with higher shrinkage pressures. Fig. 9 shows
that the rate of increase in hole size was an approximately linear
function of the shrinkage pressure, both for R = 0.1 and R = -0.3.

DISCUSSION

The results of all the tests depended strongly on the matrix shrinkage pressure. The Izod toughness increased as shrinkage pressure decreased, as predicted (6). However, the results fall into two sets, fig. 3, according to the compound used to control the shrinkage, and the work of fracture does not appear always to become indefinitely large at a particular pressure, as the theoretical equation indicates.

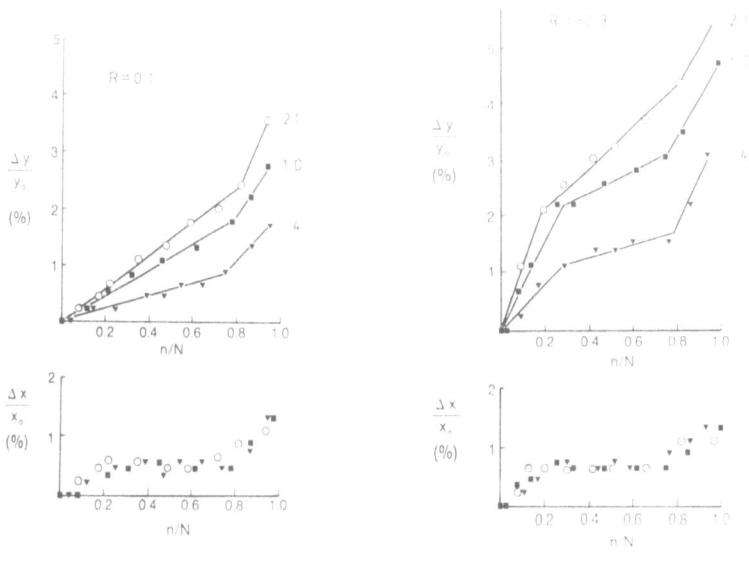

Fig. 7. Increase in hole size during fatigue at right angles to fibre (top) and in fibre direction (bottom).

Fig. 8. Increase in hole size during fatigue, at right angles to fibre (top) and in fibre direction (bottom).

These discrepancies could well be associated with another consequence of reducing the shrinkage pressure, i.e. the increase in fibre-matrix adhesion, due to more effective mechanical keying. (The mechanical keying process is most effective if the adhesive expands on cure).

Single fibre pull out experiments, carried out on production glass fibres, confirmed the reduction in shrinkage pressure observed here, but also showed much improved adhesion; so much so, in fact, that the fibres could be pulled out of the low shrinkage pressure matrices only if they were coated with silicone release agent, before being embedded in the plastic (9). There is an inverse relation between toughness and adhesion (4) (as already indicated in the Introduction). The difference between experiment and theory could be a manifestation of this.

The appearance of the fatigued specimens lends support to the role of improved adhesion. Failure took place by splitting, with the specimens having lower shrinkage pressures showing more matrix failure, and less interface failure than specimens having high shrinkage pressures, fig. 10. Also, the destruction of the adhesion by coating the fibres with silicone oil caused an increase in β, fig. 2, and an increased rate of loss of flexural modulus, fig. 6.

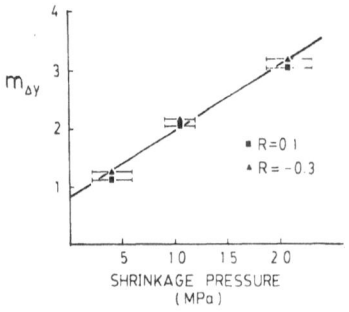

Fig. 9.   Rate of increase of hole size
vs shrinkage pressure.

The specimens with holes in them suffered damage during fatigue, by the hole opening up at right angles to the fibres. This process also, would be expected to be controlled by adhesion. Thus the specimens with reduce shrinkage pressure, and hence increased adhesion, had holes which opened up less, fig. 7 and fig. 8. These observations provided good circumstantial evidence that the effect of the additives on the fatigue properties is due, at least in part, to their promoting the improvement in adhesion between fibres and matrix.

CONCLUSIONS

The COMACO Composites show both improved impact strength

and fatigue endurance. These benefits are obtained without loss of shear or compression strength. The composites are not as tough as predicted by theory, and this is probably due to the improved adhesion between fibres and matrix. This improved adhesion appears to be an important factor in increasing the fatigue resistance of the composites.

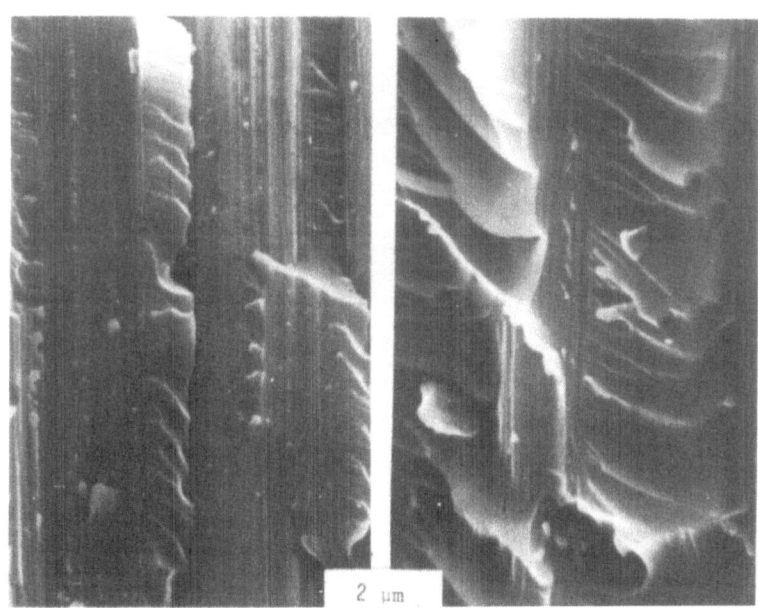

Fig. 10.   Splitting fracture surfaces of fibre composites; shrinkage pressures 21 MPa (at left) and 4 MPa (at right).

ACKNOWLEDGEMENT

The authors are grateful for the financial support received from the Department of National Defence of Canada under contract No. 08SB 97708-1-0048.

REFERENCES

1. Cooper, G.A. and A. Kelly, J. Mech. Phys. Solids 15, (1967) 279.
2. Piggott, M.R., "Load Bearing Fibre Composites", (Pergamon, Oxford), 1980.
3. Piggott, M.R., J. Mater. Sci. 16, (1970), 669.
4. Harris, B., P.W.R. Beaumont and E.M. de Ferran, J. Mater Sci. 6, (1971), 238.
5. Piggott, M.R., J. Mater Sci.16, (1981), 2837.

166

6. Piggott, M.R., Proc. ICF5, (1981), 465.
7. Lim, J.T., W.J. Bailey, M.R. Piggott, (1984) submitted to SAMPE Quarterly.
8. Piggott, M.R., U.S. Patent No. 4,427,735, (1984).
9. Chua, P.S. and M.R. Piggott, (1984) submitted to Fib. Sci. Tech.

THERMAL-MECHANICAL FATIGUE CRACK GROWTH IN INCONEL X-750

Norman Marchand* and Regis M. Pelloux**

*Research Assistant, Department of Materials Science
and Engineering, Massachusetts Institute of Technology,
Cambridge, MA 02139, USA.

**Professor, Department of Materials Science and
Engineering, Massachusetts Institute of Technology,
Cambridge, MA 02139, USA

ABSTRACT

Thermal-mechanical fatigue crack growth (TMFCG) was studied
in a γ-γ' nickel base superalloy Inconel X-750 under controlled
load amplitude in the temperature range from 300 to 650°C. In-
phase (T_{max} at σ_{max}), out-of-phase (T_{min} at σ_{max}) and isothermal
tests at 650°C were performed on single-edge notch bars under
fully reversed cyclic conditions.

A DC electrical potential method was used to measure crack
length. The electrical potential response obtained for each cycle
of a given wave form and R value yields information on crack closure and crack extension per cycle. The macroscopic crack growth
rates are reported as a function of ΔK and the relative magnitude
of the TMFCG are discussed in the light of the potential drop information and of the fractographic observations.

INTRODUCTION

Many fatigue problems in high temperature machinery such as
gas turbine components involve thermal as well as mechanical loadings. By thermal loading it is meant that the material is subjected to cyclic temperature simultaneously with cyclic stress or
strain. Analysis of the local stresses and strains versus temperature and time become very complex, consequently gross simplifications are introduced to analyse and predict the fatigue damage.
These simplifications usually involve the use of isothermal data

and life predictions techniques are based upon isothermal testing. In fact, the study of fatigue has generally bypassed real thermal fatigue loading partly because isothermal tests are relatively simple to perform, but also because it has often been felt that such tests carried out at the maximum service temperature would give worst case results. However, several studies which have compared fatigue resistance under thermal cycling conditions with that in isothermal tests have shown that in many cases, the latter, rather than giving a worst case situation, can seriously overestimate the real fatigue life [1-15].

The influence of temperature on low cycle fatigue (LCF) lives is well documented [16-19] but the mechanisms by which temperature influences the fatigue process are not well understood. Low cycle fatigue is generally acknowledged to be directly related to material ductility; however, as ductility increases with temperature, low cycle fatigue life normally decreases. Creep and environmental effects are known to influence LCF behavior, but the relative contributions of these two factors are not easily differentiated. [17-19, 20].

The subject of thermal fatigue which involves combined temperature and stress-strain cycling, is less well understood than isothermal elevated temperature fatigue and only limited data have been gathered on crack growth during thermal-mechanical fatigue (TMF) under conditions of small plastic strain [2, 5-7, 9, 13-14].

The data obtained from thermal-mechanical testing for conventionally cast Co- and Ni-based superalloys, and for directionally solidified Ni-based alloy [5-7, 9, 13] have shown faster crack growth rates than for the equivalent isothermal conditions at T_{max}. Furthermore, crack growth rates under out-of-phase cycling (T_{min} at σ_{max}) were found to be faster than under in-phase cycling (T_{max} at σ_{max}). For a 12 Cr-Mo-V-W steel thermal-mechanically cycled between 300 and 600°C, very little difference in growth rates for TMF and isothermal tests were found [21], whereas, an inverse behavior was observed on a 304 SS [15]; that is, faster crack growth rates under in-phase cycling than under out-of-phase cycling.

From the TMF crack growth data available in the open literature, it can be seen that there is no generalization to be made concerning the severity of damage associated with in-phase and out-of-phase cycling, and that there is no hard and fast rule for relating TMF data to isothermal testing.

In order to obtain a better understanding of the TMFCG behavior a research program was established with a two-fold objective. First to assess a sound, fundamental mechanistic undertanding of TMF of typical nickel-base superalloys. Second, to assess the suitability of various parameters for correlating high temperature TMFCG rates used for adequate fatigue life predictions of engine components.

EXPERIMENTAL PROCEDURE

Materials and Specimen

The material used in this investigation was a standard chemistry Inconel X-750, a corrosion and oxidation resistant material with good tensile and creep properties at elevated temperatures. The chemical composition, the heat treatment of the as-received annealed material, and the tensile properties at high and room temperature are given in Table 1. The grain size is about 0.12 mm. Single edge notch tensile bar specimens were used in this investigation. The test specimens have a rectangular cross section of 11.7 x 4.4 mm^2 and a starter notch approximately 1 mm deep which is cut by electro-discharge machining. The specimens were precracked in fatigue at 10 Hz at room temperature under a ΔK of about 10-15 MPa$\sqrt{m}$. All the ΔK's were calculated with the expression derived by Harris [22].

$$\Delta K = \Delta\sigma\sqrt{\pi a} \frac{5}{[20 - 13(\frac{a}{w}) - 7(\frac{a}{w})^2]^{1/2}} \tag{1}$$

This formula which was derived for an edge crack in a SEN plate with no bending is suitable for the testing system.

Table 1

(a) Chemical Composition (wt. pct.)

Ni	Cr	Fe	Ti	Al	Nb	Mn	Si	C	Co
72	15.5	7	2.5	0.7	1	0.5	0.2	0.04	1.0

(b) Heat Treatment

Temperature (°C)	Time (hr)	Quench
1150	2	air cool
850	24	air cool
700	20	air cool

(c) Tensile Properties

Temperature (°C)	σ_y (MPa)	σ_{UTS} (MPa)	ε_p (Pct)
24	610	1000	30
300	616	1000	32
650	550	820	7

170

Apparatus and Test Conditions

The apparatus used in this study was a computer-controlled thermal fatigue testing system which consisted of a closed-loop servo-controlled, electro-hydraulic tension-compression fatigue machine, a high frequency oscillator for induction heating, an air compressor for cooling, a mini-computer, etc. Figure 1 shows the control block diagram of this system.

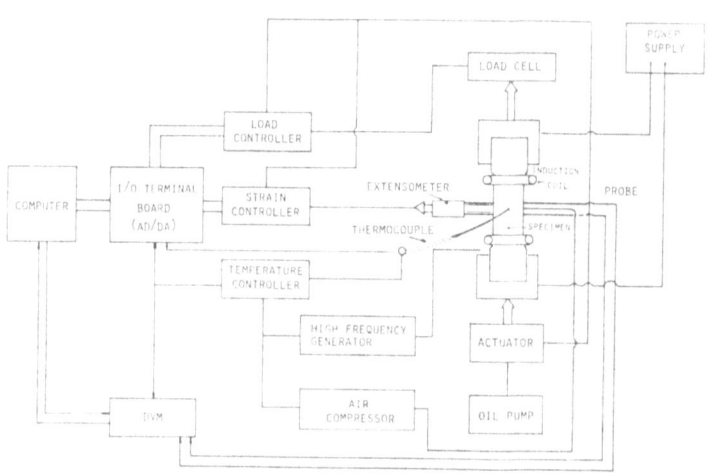

Fig. 1. Diagram showing the control system of apparatus used for TMFCG tests.

The system is capable of testing specimens of different sizes and configuration (SEN, CT, hollow tube, etc.) up to loads of 25,000 lbs. The specimen alignment is insured by the use of a Wood's metal pot which prevents a bending moment in the specimen. Because a DC potential drop technique is used to monitor crack growth, the lower grip is electrically insulated from the system by means of a ceramic coating. The ends of the grips are water cooled by means of copper coils.

Temperature was measured with 0.2 mm diameter chromel-alumel thermocouples which were spot welded along the gauge length. By computer controlling, the temperature in the gauge length was maintained within 5°C of the desired temperature for both axial and transverse directions over the entire period of the test.

Temperature and stress were computer controlled with the output of the thermocouple so that they were in-phase or out-of-phase for the same triangular wave shape. Therefore, specimens were in tension at low temperature and in compression at high temperature under the out-of-phase cycling, and vice versa under the in-phase fatigue. The temperature range in these tests was 300 to 650°C. The tests were carried out at a frequency of 0.0056 Hz (1/3 cpm) and were run at a R ratio ($\sigma_{min}/\sigma_{max}$) of -1 or 0.05. Isothermal fatigue tests were also conducted under the same frequency at T_{max} for comparison with the results of TMFCG tests. All the tests were carried out in air. Table 2 summarizes the experimental conditions. At least two tests at each condition were performed to insure repeatability of the results.

There have been few attempts to measure crack length in the TMF cycling using the potential drop technique [2]. The method has proved to be satisfactory in isothermal conditions and can be used for TMF testing provided that the electrical noise is adequately filtered and the calibration curve is properly corrected to take into account changes of potential with temperature. This was achieved by using a 450 KHz filter and a high accuracy digital voltmeter programmed to convert the analog signal average over 10 power line cycles (PLC). In this mode, 1 PLC is used for the run-up time with the A/D conversion repeated ten times. The resulting ten readings are then averaged and the answer becomes a single reading.

This system, as opposed to the optical measurement and compliance methods, has the capability of monitoring the crack extension during a single cycle, whereas the previous methods are limited for growth measurement to no more than one cycle. Therefore, detailed analysis of the crack growth process can be performed and this is particularly important in trying to determine the mechanisms involved with TMFCG or with the kinetics of short crack growth.

Table 2

Experimental Conditions

Frequency (Hz)	Temperature ($^\circ$C)	R-Ratio
0.1	25	0.05
0.0056	650	0.05
0.0056	650	-1
0.0056	300-650 (in-phase)	0.05
0.0056	300-650 (in-phase)	-1
0.0056	650-300 (out-of-phase)	-1

Results and Discussion

Figures 2 and 3 show the variation of the potential with

simultaneous change of the net stress and temperature for in-phase and out-of-phase cycling at low and high ΔK. In these figures, the potential change with stress and temperature $V(T,\sigma)$ and with temperature only $V(T)$ are plotted. The mechanical driving force $V(\sigma)$ which is the difference between $V(T,\sigma)$ and $V(T)$ are also plotted. In order to assess crack growth, the peak potential of each cycle was recorded and plots of the voltage versus the number of cycle (N) were obtained. Using an experimental calibration curve, the crack lengths versus N were derived. The crack growth rates were calculated using a seven-point incremental polynomial method. Figure 4 shows the results as a function of ΔK (R = 0.05) and K_{max} (R = -1).

First, it can be seen that the crack growth rates are higher for TMF cycling than for the equivalent isothermal condition (650°C) which is in agreement with the results obtained on other nickel-base alloys [5-7, 9, 13]. Secondly, it is observed (Fig. 4) that the crack growth rates are higher for R = - 1 than for R = 0.05 which indicates that compressive stresses play an important role in the mechanics of TMFCG. Comparison between out-of-phase and in-phase at R = -1 shows that out-of-phase cycling is more damaging than in-phase cycling at high ΔK, whereas at low ΔK, the crack growth rates are the same. The explanation for this behavior can be found in Figs. 2 and 3. At low ΔK the potential curves $V(\sigma)$ for out-of-phase and in-phase are similar which indicates that the mechanical driving force for cracking are similar and identical crack growth rates are therefore expected. As ΔK increases, however, the $V(\sigma)$ potential curves for both in-phase and out-of-phase cycling display characteristic features (Fig. 3). The in-phase $V(\sigma)$ curve shows a smooth increase with σ_{net} up to the maximum followed by a sharp increase at σ_{max}. The potential then remains stable as σ_{net} starts to decrease and sharply falls as σ_{net} approaches zero. In the compression regime the potential smoothly decreases, reaches a minimum at σ_{min}, and finally increases as the stress increases again. On the other hand, the out-of-phase $V(\sigma)$ potential curves shows (at the same ΔK) a smooth increase with σ_{net} with a peak value at σ_{max}. The potential then decreases down to a minimum value and finally increases again with σ_{net}. The most important feature of these signals is the crossover point (denoted A) for which $V(\sigma) = 0$. The crossover point represents the stress to apply to the specimen for the potential to equal $V(T)$. Because $V(T)$ is measured at $\sigma = 0$ for the entire thermal cycle, the expected stress to apply is $\sigma = 0$. However, if the potential field near the crack tip is disturbed either by a non-zero residual stress-strain field or by geometrical events such as blunting, closure, etc., the $V(T,\sigma = 0)$ potential might not necessarily equal $V(T)$ and a non-zero stress is required to cancel out the contribution of this phenomena. For the out-of-phase potential, the crossover occurs at a negative stress ($\sigma = -75$ MPa at $\Delta K = 50$ MPa$\sqrt{m}$), whereas, it always occurs at zero stress for the in-phase potential. By

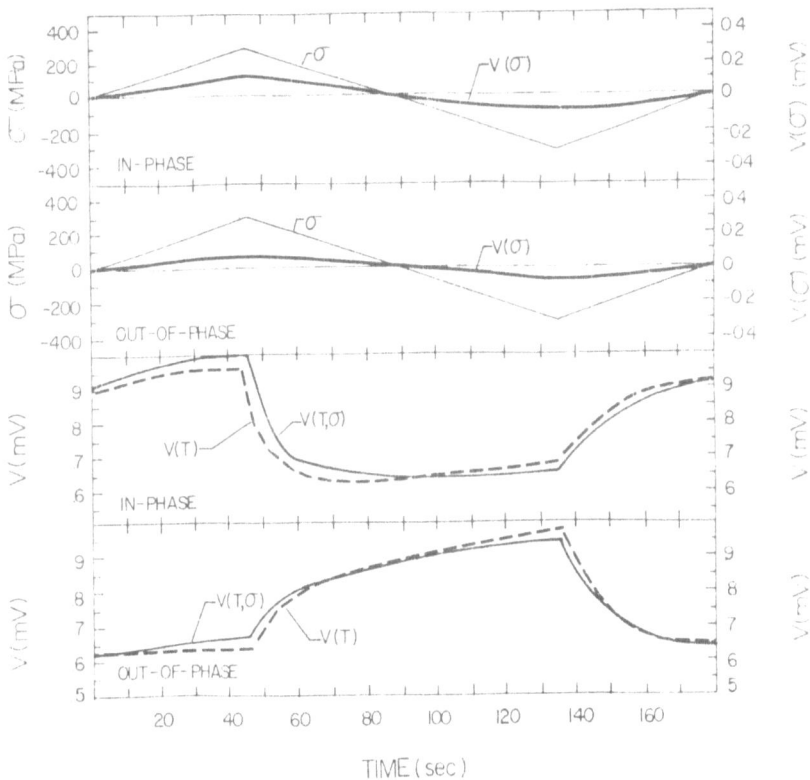

Fig. 2. Potential curves $V(T,\sigma)$, $V(T)$ and $V(\sigma)$ for TMFCG of Inconel X-750. $\Delta K = 25$ MPa$\sqrt{m}$.

first assuming that the crossover point occurs near an effective closure stress (σ_{cl}) it follows that the effective stress intensity factor (ΔK_{eff}) for crack growth will be higher for out-of-phase than for in-phase cycling (σ_{cl} is negative). If we plot the crack growth rates as a function of ΔK_{eff}, we find that both in-phase and out-of-phase crack growth rates overlap (Fig. 5).

It is important to note that the absolute amplitude of the $V(\sigma)$ signal (see Fig. 3) in the compressive regime of the cycle, is much higher for out-of-phase than for in-phase cycling which indicates that the crack surfaces are in contact on a much larger scale than under in-phase cycling. This was confirmed by fractographic observation which shows that out-of-phase cycling leads

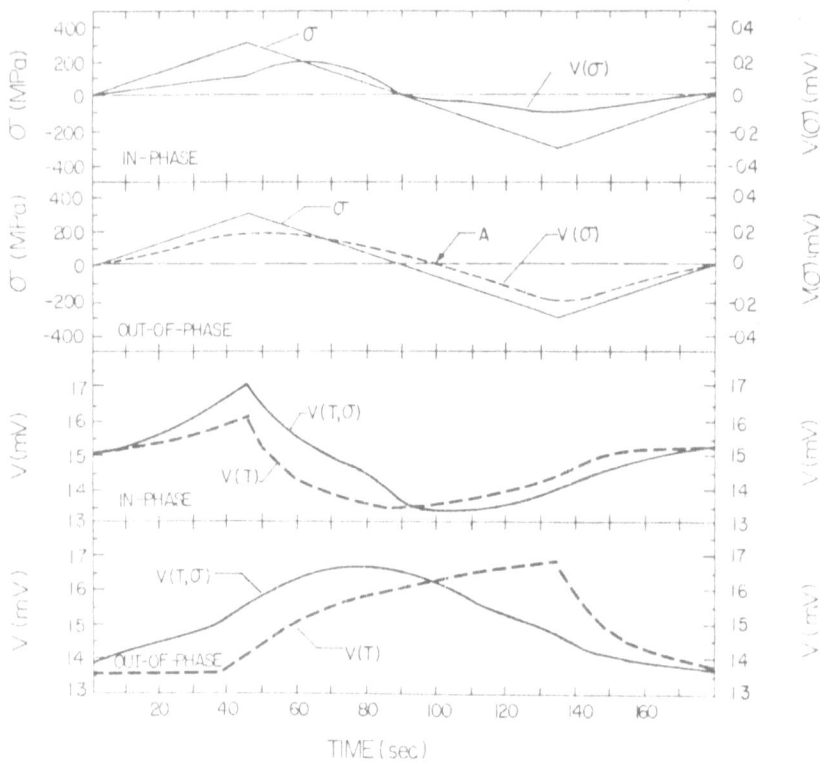

Fig. 3. Potential curves $V(T,\sigma)$, $V(T)$ and $V(\sigma)$ for TMFCG of Inconel X-750. $\Delta K = 50$ MPa$\sqrt{m}$.

to transgranular cracking with considerable mating, whereas, in-phase cycling leads to intergranular fracture with little evidence of mating. The higher crack growth rates at R = -1 than R = 0.05 and the fractographic observations lead to the conclusion that although there should be no cracking at the maximum temperature because of the compressive stress, some form of severe damage is taking place under compressive strain. Also, the surface oxide film formed at high temperature will rupture at low temperature, under maximum tensile stress. This leads to a resharpening of the crack tip with each cycle and results in an increase in growth rate.

The rationale for the negative closure stress in out-of-phase cycling can be found by assuming that the residual stress field at the crack tip is of tensile nature at zero applied stress and that a negative stress has to be applied in order to cancel it and to close the crack. This negative σ_{c1} also represents the effective contribution to cracking of the damage taking place during the compression part of the cycle.

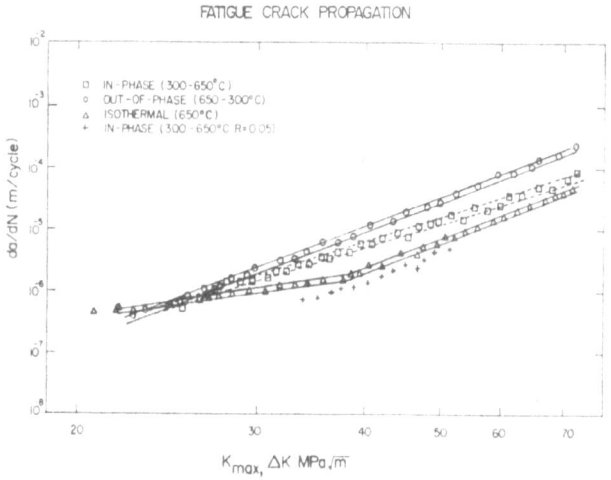

Fig. 4. TMFCG of Inconel X-750 as a function of ΔK.

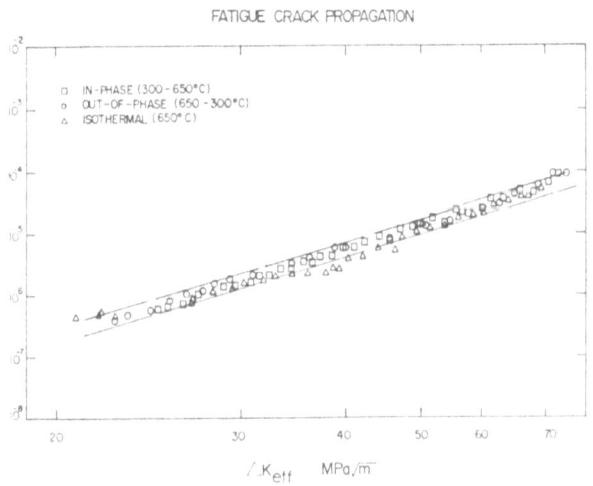

Fig. 5. TMFCG of Inconel X-750 as a function of ΔK_{eff}

176

The difference in growth rates between the isothermal test (650°C) and the TMF tests were also explained by looking at their respective $V(\sigma)$ potential curves at low and high K. At low ΔK no significant differences were observed between $V(\sigma)$ potential curves. At higher ΔK, however, the isothermal $V(\sigma)$ potential has shown a crossover point taking place at a positive stress. This has the effect of reducing the effective driving force for cracking. By taking into account this σ_{cl} in the computation of ΔK, one finds that the crack growth rates in terms of ΔK_{eff} for both isothermal and TMF tests are similar (Fig. 5). The positive closure stress observed was attributed to the build-up of oxides in the wake of the crack. This is supported by fractographic observation which shows greater oxidation for isothermal testing than for TMF testing.

CONCLUSIONS

Faster crack growth rates were measured in Inconel X-750 cycled between 360 and 650°C under out-of-phase conditions than under in-phase cycling at R = -1 or R = 0.05. This behavior was rationalized by introducing the concept of an effective closure stress which was defined as the applied stress at the crossover point of the $V(\sigma)$ potential curve. σ_{cl} should represent the effective contribution to cracking of the damage taking place during the compressive part of the cycle. Correlation between TMFCG rates and ΔK appear to be valid provided elastic conditions prevailed in the bulk. Correction to the applied stress can be introduced in order to take into account damage occurring in the compressive part of the last cycle. Further work is needed to assess the suitability of ΔK to more realistic conditions involving TMFCG under cyclic plastic strains.

Acknowledgment: This research work was sponsored by the National Aeronautics and Space Administration under Grant No. NAG3-280.

REFERENCES

1. Sheffler, K. D., Vacuum Thermal-Mechanical Fatigue Behavior of Two Iron-Base Alloys, ASTM STP 612, (1976) 214-226.

2. Skelton, R. P., Environmental Crack Growth in 1/2Cr-Mo-V steel during Isothermal High Strain Fatigue and Temperature Cycling, Mat. Sci. Eng., vol. 2, (1978) 287-298.

3. Fujino, N. and Tarai, S., Effect of Thermal Cycle on Low Cycle Fatigue Life of Steels and Grain Boundary Sliding Character- istics, ICM 3, vol. 2, 1979, 49-58.

4. Kuwabara, K. and Nitta, A., Thermal-Mechanical Low Cycle Fatigue under Creep-Fatigue Interaction in Type 304 Stainless Steel, ICM 3, vol. 2, 1979, 69-78.

5. Rau, C.A., Gemma,A.E.,Leverant, G.R., Thermal-Mechanical Fat- igue Crack Propagation in Nickel and Cobalt Base Superalloys under Various Strain-Temperature Cycles, ASTM STP 520 (1973) 166-178.

6. Gemma, A.E., Langer, B.S. and Leverant, G.R., Thermal-Mech- anical Fatigue Crack Propagation in Anisotropic Nickel Base Superalloys", ASTM STP 612 (1976) 199-213.

7. Gemma,A.E., Ashland, F.X. and Masci, R.M., The Effects of Stress Dwells and Varying Mean Strain on Crack Growth During Thermal Mechanical Fatigue, J. Test. Eval., vol. 9, no.4, (1981) 209-215.

8. Troshchenko, V.T. and Zaslotskaya, L.A., Fatigue Strength of Superalloys Subjected to Combined Mechanical and Thermal Loading, ICM 3, vol. 2, (1979) 3-12.

9. Meyers, G. J., Fracture Mechanics Criteria for Turbine Engines and Hot Section Components, NASA CR-167896, (1982).

10. Jaske, C.E., Thermal Mechanical Low Cycle Fatigue of AISI 1010 Steel, ASTM STP 612, (1976) 170-198.

11. Westwood, H. J. and Moles, M.D., Creep-Fatigue Problems in Electricity Generation Plants, Can. Met. Quart., vol.18, (1979) 215-230.

12. Bhongbhobhat, S., The Effect of Simultaneously Alternating Temperature and Hold Time in the Low Cycle Fatigue Behavior of Steels, Low Cycle Fatigue Strength and Elasto-Plastic Behavior of Materials, eds. Rie, K.T. and Harbach, E., DVM (1979), 73-82.

13. Leverant, G. R., Strongman, T.E., and Langer, B.S., Parameters Controlling the Thermal Fatigue Properties of Conventional Cast and Directionally-Solidified Turbine Alloys, Superalloys: Metallurgy and Manufactures, ed., Kear, B.H., Muzuka, D.R., Tien, J.K., and Wlodek, S.T., Claxtors Pub.(1976) 285-295.

14. Okazaki, M. and Koizumi, T., Crack Propagation During Low Cycle Thermal-Mechanical and Isothermal Fatigue at Elevated Temperatures, Met. Trans. vol. 14A, (1983) 1641-1648.

15. Kuwabara, K., Nitta, A, and Kitamura, T., Thermal-Mechanical Fatigue Life Prediction in High Temperature Component Materials for Power Plants, Conf. on Advances in Life Prediction Methods, ASME-MPC, Albany, N.Y. (1983) 131-141.

16. Wareing, J., Mechanisms of High Temperature Fatigue and Creep-Fatigue Failure in Engineering Materials, Fatigue at High Temperature, ed. Skelton, R.P., Applied Sci. Pub., (1983) 135-185.

17. Lloyd, G. J. , High Temperature Fatigue and Creep-Fatigue Crack Propagation: Mechanics, Mechanisms and Observed Behavir in Structural Materials, Fatigue at High Temperature, ed. Skelton, R.P., Applied Sci. Pub. (1983) 187-258.

18. Coffin, L. F., Damage Processes in Time Dependent Fatigue - A Review, Creep-Fatigue-Environment Interactions, eds. Pelloux, R. M. and Stoloff, N.S., AIME (1980) 1-23.

19. Mills, W. J. and James, L.A., Effect of Temperature on the Fatigue Crack Propagation Behavior of Inconel X-750, Fat. Eng. Mat. Struct., vol. 3, (1980) 159-175.

20. Skelton, R.P., The Growth of Short Cracks During High Strain Fatigue and Thermal Cycling, ASTM STP 770, (1982) 337-381.

21. Koizumi, T. and Okazaki, M., Crack Growth and Prediction of Endurance in Thermal-Mechanical Fatigue of 12 Cr-Mo-V-W Steel, Fat. Eng. Mat. Struct., vol. 1, (1979) 509-520.

22. Harris, D. O., Stress Intensity Factors for Hollow Circumferentially Notched Round Bars, J. Basic Eng., vol. 89 (1967) 49-54.

MICROSTRUCTURAL FEATURES OF HIGH-TEMPERATURE LOW-CYCLE FATIGUE FAILURE OF MAR-M-200(+Hf) ALLOY UNDER BASIC STRAINRANGE PARTITIONING MODES OF STRAINING

S. Nadiv and A. Berkovits
Dept. of Materials and Dept. of Aeronautical Engineering
Technion-Israel Institute of Technology, Haifa.

ABSTRACT

Cyclic loading tests, under four different strainrange partitioning modes, on PWA 1422 (MAR-M-200+Hf) alloy at 975°C were conducted to gain understanding of the microstructural features of failure responsible for observed deviations from theoretical predictions. Time-independent plasticity and time-dependent creep deformations were obtained by means of different strain-rates controlled through different cyclic loading frequencies. Scanning electron metallography and fractography revealed the important role of the heterogeneous microstructural features of the dendritic unidirectionally solidified material, the importance of relaxation phenomena such as the Ostwald ripening of the gamma-prime particles, and of oxidation, on the low-cycle fatigue crack propagation. The cyclic response was compared to the monotonic response of the material through a material hodograph presentation.

INTRODUCTION

High-temperature low-cycle fatigue life of nickel-base superalloys can be predicted by the strainrange partitioning procedure[1]. It involves the determination of four basic cyclic hysteresis loops which combine two types of inelastic strains with two loading directions within a predetermined strainrange value, namely: PP-plastic tension and plastic compression, CC-tensile creep and compression creep, PC-plastic tension and compression creep, and CP-tensile creep and plastic compression. The different modes can be achieved by controlling either temperature or strain-rate. High strain-rate and/or relatively low temperature induce the mode of plasticity (time-independent loading), whereas low strain-rate at high temperature induces the creep mode (time-dependent loading).

The strain-rate within a given strainrange dictates the maximum attainable stress in the strength test mode, whereas the stress level in the creep mode controls the creep duration. The cyclic response of the material is therefore a function of the inelastic strainrange, the strain-rate and the cyclic stress, controlled by the wave-form and the frequency of cyclic straining[2].

MATERIAL CHARACTERISTICS

The study concentrates on a dendritic unidirectional solidified MAR-M-200(+Hf) nickel-base superalloy known as PWA 1422. According to Manson[3] the fatigue life of this material is insensitive to the mode of applied strainrange. However, in the present work some deviations in life were observed, probably due to nonuniform microstructural features. The commonly used dendritic unidirectionally solidified material differs from idealized cellular unidirectionally solidified material by having dendritic-arm protrusions along the longitudinal grain-boundaries, so that segments of grain-boundaries are normal to the solidification direction, which is the loading direction in service (Fig. 1).

The microstructure within the dendritic arm shows fine eutectic-like nodular structure of γ'-phase, which coarsen towards the arm periphery with the appearance of a more condensed grainy form (Fig. 2). It differs from the usually observed very fine modulated square γ'-particles coherent with the γ-phase matrix, which appear in the grain bulk material (Fig. 3). The grain-boundaries are heavily decorated with a mixture of γ'-particles and $M_{23}C_6$ carbides. Occasionally, large MC carbides or TiN-nitrides appear adjacent to the dendritic-arm tips.

EXPERIMENTAL PROGRAM

Mechanical tests[4] were conducted in a 25-ton closed-loop electrodydraulic MTS machine, controlled by integral function generator. Differentiation between loading and reversal rates was obtained with an Interstate Electronic function generator equipped with suitable dual-ramp facility. Strain was measured with an extensometer having two strain transfer rods, each attached at one end to gage-length collars on the specimen, and at the other to an LVDT outside the furnace. The averaged signal

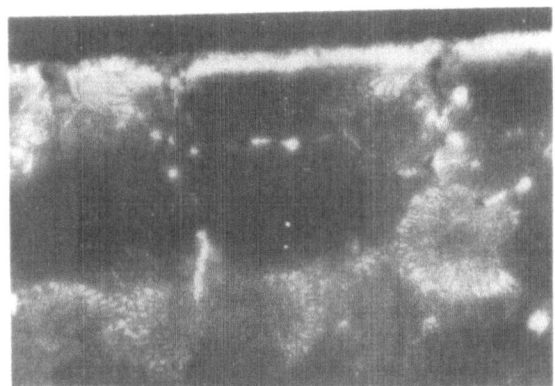

Fig. 1. Dendritic Unidirectionally Solidified Structure of
PWA 1422 (X300).

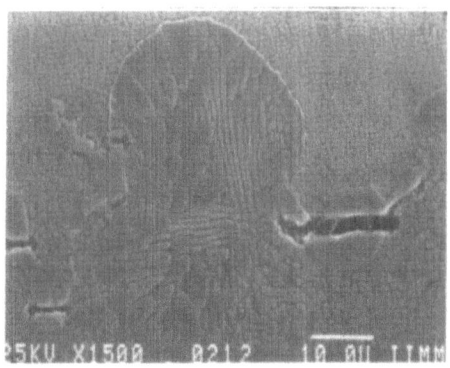

Fig. 2. Dendritic-Arm Structure Showing Microcrack Initiation in
Monotonic Creep (X1500).

Fig. 3. Modulated Square γ'-Particles Coherentwith Bulk Matrix
(X7500).

was fed into the strain control function of the testing machine, and together with load-cell measurements recorded as related functions, or functions of time on an X-Y recorder. An Amsler tube furnace was controlled in a closed-loop mode using a Thermac electronic controller to keep constant temperature within $\pm 1°C$ for long periods. Typical hysteresis stress-strain reversal loops for strainrange of 0.011 used in the present study at 975°C are shown in Fig. 4.

Following testing metallographic and fractographic observations were made with JEOL C-35 or T-200 scanning electron microscopes on samples polished and etched with the following reagents: 25 gr. $CuSO_4$ dissolved in a mixture of 100cc HCl, 12.5cc H_2SO_4, and 112.5cc water. Under optical microscopy the metallographs appear tinted with feature-differentiating colors. All SEM specimens obtained gold shadowing before examination.

RESULTS AND INTERPRETATIONS

The results of all SRP tests, presented as strainrange versus cycles-to-failure curves, are shown in Fig. 5. The PP-test results show significantly larger LCF lives than the CC, CP, and PC test results. The latter, which have creep-mode segments in the hysteresis loop, have similar LCF lives. A typical sequence of recorded CC-type hysteresis loops in the course of testing time is shown in Fig. 6. The crack initiation and development is manifested by a continuous drop in the tensile stress amplitude of each consecutive loop. The inelastic true strain, within the controlled strainrange (shown by the straightness of the loop), decreases as well, as the crack-opening-displacement (shown by the bending of the upper loop part) increases towards failure. A typical stress-amplitude history is shown in Fig. 7. In all cases containing creep mode significant crack development is noticed during the last 25 percent of the cyclic life. The stress amplitudes developed within the selected strainrange were well below the ultimate stress observed in monotonic tests at the same temperature and strain-rate. In the monotonic stress-strain curve, upon straining beyond the ultimate stress, dynamic recovery was noticed for true strains over 0.1 before fracture occurred[5].

Metallographically it was observed that during the recovery dissolution of the fine modulated γ'-particles takes place; the mean-free-distance between the longitudinal rows of particles increases as a result of their size diminution (Fig. 8). This is believed to act as a means of stress relaxation, even under lower stress loadings at points of high stress concentration. In monotonic creep tests under stress levels below the ultimate

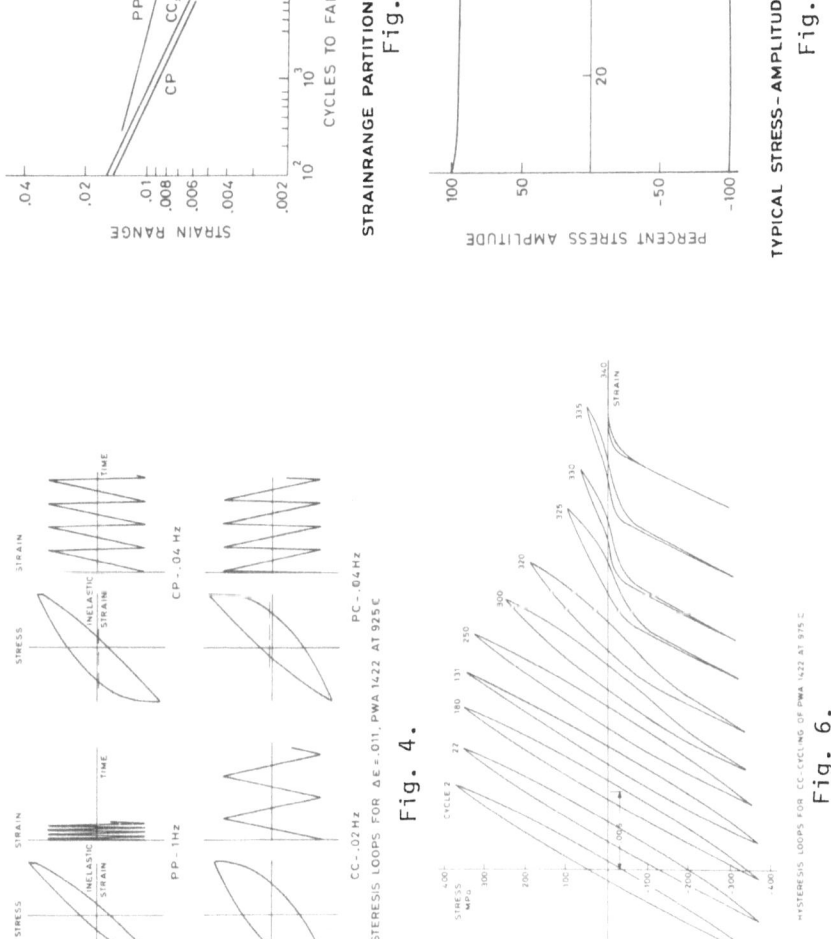

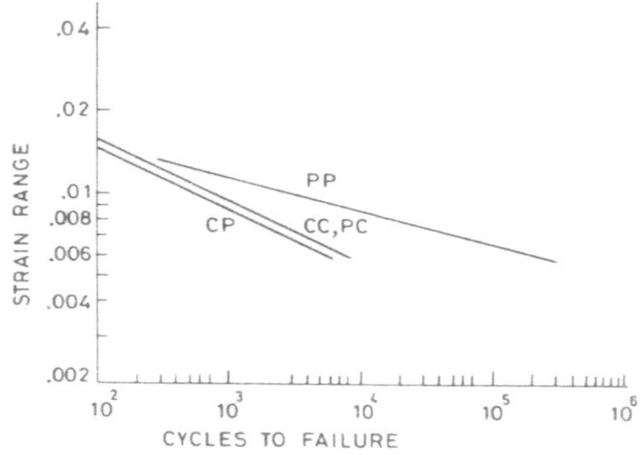

STRAINRANGE PARTITIONED LIFE OF PWA 1422 AT 975 C

Fig. 5.

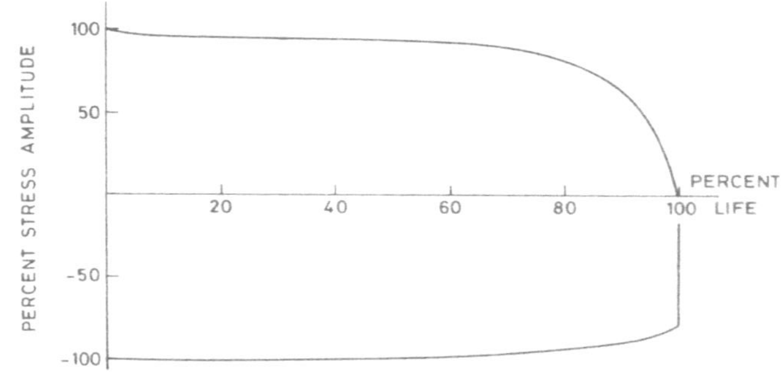

TYPICAL STRESS-AMPLITUDE HISTORY UNDER STRAIN CONTROL

Fig. 7.

HYSTERESIS LOOPS FOR Δε=.011, PWA 1422 AT 925 C

Fig. 4.

HYSTERESIS LOOPS FOR CC-CYCLING OF PWA 1422 AT 975 C

Fig. 6.

value, the relaxation was found to take place in a different way, Ostwald ripening of the fine γ'-particles, in places of high microstress concentrations, such as adjacent to large MC carbides, or dendritic-arm protrusions (Fig. 9). The coherent γ'-particles grow into large and irregular-form massive particles.

Under cyclic loading, in all SRP mode tests, all fractures appeared to be intergranular. A single crack propagated along the grain-boundary segments which surround the dendritic-arm protrusions, and through zones of inter-dendritic micro-segregation within the unidirectionally solidified grains where coarse γ'-particles exist before crack propagation. In a few instances short cracks branching away from the main propagating crack were seen. Otherwise no crack initiation other than the propagating crack was detected, probably due to the fact that the bulk material elsewhere is well relaxed. This finding contradicts observations in monotonic loaded specimens, in both plasticity or creep regimes, where many cracks were observed to initiate in many different localities, near MC carbides or on grain-boundaries around the dendritic-arm protrusion (Fig. 2), later to coalesce slowly under further straining and to lead to fracture of partially intergranular appearance. Transcrystalline segments are more readily observed, by the appearance of dimples, or sometimes of serrated cleaved fractures, uprooted in a curious block-like appearance of surfaces normal to each other (Fig. 10). Occasionally, nodular appearance of γ'-particles can be seen in the fractographs (Fig. 11). In some cases of monotonic or cyclic loading fracture occurred at the end of the specimen gage-length, adjacent to the collars, as the alloy in question is very sensitive to local constraints (tapered triaxiality, stress concentrations, etc.).

In all SRP tests, apart from the PP-test, the fracture surfaces were heavily covered with oxide film (Fig. 12), since the tests were conducted under air. Oxidation is a decisive factor in LCF crack propagation[6], although most of it took place during the tensile crack-opening-displacement upon the already cracked surfaces. Beneath the oxide film oxygen diffused into the material to cause the appearance of new continuous phase parallel with the exposed surfaces. It seems that depletion of active elements, such as chromium, tungsten, and cobalt occurs as these elements diffuse towards the oxide film. The fine γ'-particles completely disappear within it. Below this phase ripened γ'-particles are observable. The heavy oxide film eliminated possible observation of fatigue crack striations, but the general appearance of the fracture was conserved.

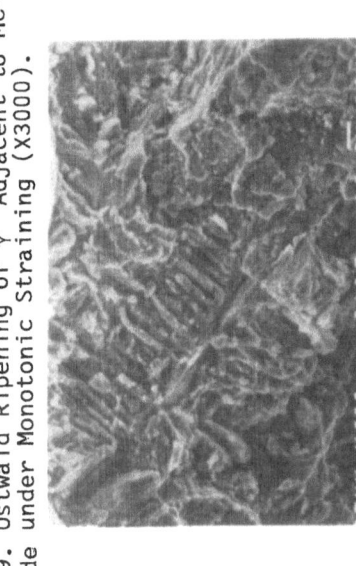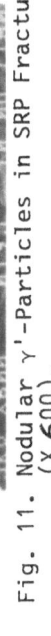

Fig. 9. Ostwald Ripening of γ' Adjacent to MC-Carbide under Monotonic Straining (X3000).

Fig. 11. Nodular γ'-Particles in SRP Fracture (X 600).

Fig. 8. Dissolution of γ' During Dynamic Recovery under Monotonic Straining (X7500).

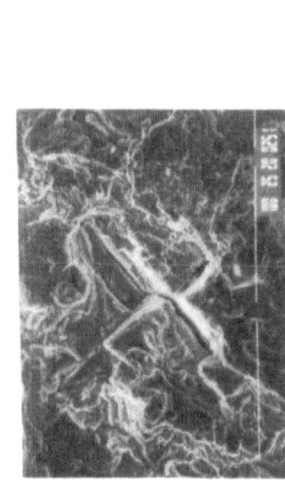

Fig. 10. Serrated-Cleavage Transcrystalline Fracture under Cycling Strain (X3000).

COMPARISON BETWEEN MONOTONIC AND CYCLIC LOADING RESPONSE

The material hodograph[5,7], based on monotonic loading curves for both plasticity or creep tests, was used as a framework for comparison (Fig. 13). The coordinates are inelastic strain-rate versus true inelastic strain, with true stress as a discrete parameter. It serves both tension and compression tests. Horizontal lines describe monotonic strength tests, whereas constant stress contour lines describe monotonic creep tests. The line connecting all stress minima describes the ultimate stress situation in strength tests, or the secondary creep stage in constant-stress creep. The dividing strain-rate value between creep and plasticity conditions is about 0.05 min^{-1}.

For the cyclic loading hodograph, the hysteresis loop stresses and inelastic strains have been halved to enable comparison with the monotonic data. The cyclic response of the material depends much on the straining frequency. All creep segments of the hysteresis loops in the CC, CP, and PC tests fall on a single straight line in the creep region of the hodograph, and the stress level at any combination of creep strain-rate and strain, invariant to the overall stress amplitude, lies on a line extended tangentially from a constant monotonic stress contour curve at low strain values. This means that in order to develop a given stress at a given inelastic strain, a higher strain-rate is needed under cyclic conditions than under monotonic conditions. At higher frequencies the common line for all creep segments of the different hysteresis loop modes moves to higher strain-rates and stress levels, as the available time for operating the creep mechanisms becomes shorter. The constant stress (tangent) line of the hodograph for cyclic loading shows that the transient creep stage, signified by the reduction of strain-rate under monotonic loading, is avoided. Apparently, at half-stress amplitude of 300 MPa, no strain-rate hardening takes place at all. At higher amplitudes some hardening is noticed, while at lower amplitudes dynamic recovery sets in at very low strains. The factors which are believed to contribute most to the unique cyclic creep behavior of the PWA 1422 alloy, result from the dynamic recovery, as well as the fact that the inelastic straining results from localised microstructural changes, adjacent to the dendritic-arm protrusions, with almost no grain-boundary sliding[2]. Thus the same unique creep mechanism controls all cyclic creep behavior of the material before crack initiation.

Similar straight line results were obtained in the plasticity region of the hodograph, from the plasticity segments of the PP, PC, and CP tests. However, correlation with the monotonic loading hodograph has not yet been completed. As high strain-rates had to be used for obtaining the inelastic plasticity behavior, there was a considerable delay in the flow response of the material immediately after strain reversal[8] and the continuous transient behavior may complicate correlation. In the hodograph for cyclic loading, the inelastic strain is uniform along the gage-length of the specimen, up to the ultimate-conditions dividing line. Beyond it the situation is localised to the tip of the propagating crack as failure starts, with the inelastic strain actually decreasing as crack-opening-displacement increases with each consecutive loop until fracture occurs.

ACKNOWLEDGMENT

The authors are grateful to Mr. G. Shalev for assistance in the scanning electron microscopy, and to Mr. S. Nahmani and Mr. A. Grunwald for help in the mechanical tests laboratory.

REFERENCES

1. Manson, S.S., Halford, G.R. and Hirschberg, M.H., Creep-Fatigue Analysis by Strainrange Partitioning, Proc. 1st. Symp. on Design for Elevated Temperature Environment, A.S.M.E., San Francisco, 12-28 (1971).

2. Sidey, D. and Coffin, Jr., L.F., in Fatigue Mechanisms, (ASTM STP No. 675) 528-568 (1979).

3. Manson, S.S. and Halford, G.R., Complexities of High Temperature Metal Fatigue: Some Steps Toward Understanding, Proc. 25th Israel Annual Conf. on Aviation and Astronautics, Haifa (1983).

4. Berkovits, A. and Nadiv, S., Creep-Fatigue in PWA 1422 Material, Proc. 5th Internat. Congress of Experimental Mechanics, Montreal (1984).

5. Berkovits, A. and Nadiv, S., Constitutive Relationships for Creep-Fatigue in High-Temperature Materials, Proc. Mechanical Behavior of Materials - IV, 1, 149-156 (1983).

6. Runkle, J.C. and Pelloux, R.M., in Fatigue Mechanisms, (ASTM STP 675) 501-524 (1979).

7a. Berkovits, A., Hodographic Approach to Predicting Inelastic Strain at High Temperature , NASA TN D-6973 (1972).

188

7b. Berkovits, A., Prediction of Inelastic High Temperature
    Materials Behavior by Strain-Rate Approach , ASME J. of
    Engineering Materials and Technology, 96H, 106-108 (1974).

8. Rosen, A., Delay Time During Creep of Aluminium , Inst.
   Metals, 99, 111-114 (1971).

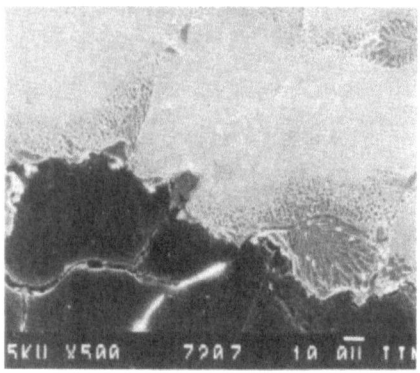 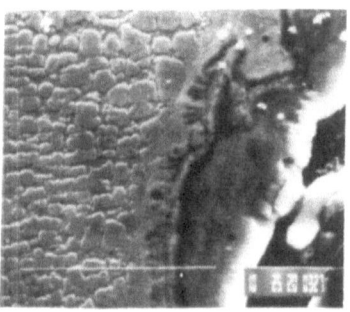

a) Through Interdendritic Micro- b) Oxide Film, Oxidized Single-
   Segregation Zones and around   Phase Layer, and Ripened
   Dendritic Arms.   γ'-Particles.

Fig. 12. Oxidation at SRP Fracture Surface.

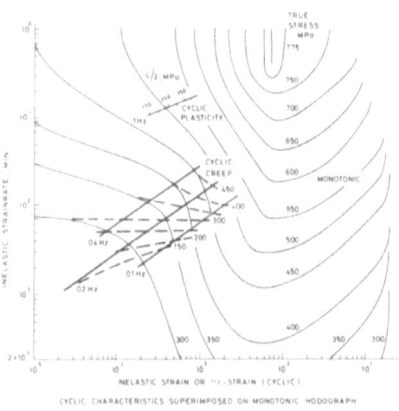

Fig. 13.

ON CERTAIN FRACTURE MECHANICS CONSIDERATIONS IN ENVIRONMENTAL CRACKING

A. Neimitz[1]   and   E.C. Aifantis

Department of Mechanical Engineering and Engineering Mechanics
Michigan Technological University, Houghton, MI 49931

ABSTRACT

A combination of macroscopic fracture mechanics and microscopic defect physics considerations is adopted to interpret hydrogen-controlled cracking. A distinction is made between the "plastic zone" which defines the structure of the stress field ahead of the crack tip due to external loading and the "process zone" which accomodates the hydrogen species and defines the distribution of the cohesive forces in the immediate vicinity of the crack controlling fracture. The crack kinetics is modelled on a discrete basis in terms of an average critical concentration of hydrogen in the process zone and an average critical time for hydrogen species to travel the process zone distance and build up its concentration.

INTRODUCTION

Environmental cracking is a challenging problem of modern technology revealing the need of adopting an interdisciplinary approach based on both macroscopic fracture considerations and microscopic physical modelling of the degradation process occuring in the vicinity of the crack tip. The problem has been put within a unifying chemomechanical framework for general degradation processes by Aifantis [1]. Unger and Aifantis [2] and Unger, Gerberich and Aifantis [3] have concentrated on elastodiffusion effects at the crack tip and established simple relationships between crack velocities, external pressures, and stress intensity factors which compared sufficiently well with certain experimental data.

---

[1] On leave from Technical University of Kielce, Kielce, Poland

In the above work the stress distribution due to the external loading was assumed to be elastic. The degradation process occuring ahead of the crack tip was assumed to be a diffusion-like accumulation of hydrogen assisted by both concentration and stress gradients. The concentration distribution was obtained from the solution of a partial differential equation modelling coupled stress-solute transport effects [1]. Finally, the velocity of the crack was postulated to be a linear function of the concentration at a critical distance (which could be identified with the length of the process zone) ahead of the crack tip. This model, however simple, has produced realistic relations between pressures and stress intensity factors (log P - K_I, for threshold conditions) as well as velocities and stress intensity factors (log V - K_I, for kinetic conditions) for some ceramics and high strength steels. Moreover, the values of the phenomenolopical constants determined from the log P - K_I graphs compare well to those determined from the log V - K_I graphs. Experimental data of Oriani et. al., Evans, Wiederhorn et. al., Gerberich et. al., etc. are interpreted very well by the model as discussed in [3].

In the present work we retain the same philosophy as in the above model but we explore other alternatives concerning the stress distribution due to external loading and the nature of the process zone; the critical concentration of hydrogen within the process zone and the extra strains induced on the lattice; and the velocity of the crack calculated as the ratio of an average jump of the crack (which turns out to be a portion of the length of the process zone or the process zone itself) over the time interval required to build up a critical concentration within the process zone.

The structure of the stress field due to the external loading is assumed to be plastic and our conclusions here are motivated by the nonlinear inelastic fracture mechanics analyses of Hutchinson [4] for sharp cracks and Rice and Johnson [5] for blunted cracks. Indeed, we propose a method analogous to Barenblatt's [6] to eliminate the plastic singularity in [4] and then in conjunction with the results of [5] we can conclude the length of the process zone and the distribution of cohesive forces and displacements within it.

The critical concentration of hydrogen within the process zone is calculated in terms of a critical displacement induced by hydrogen dissolved in the lattice which superimposes on the lattice displacement induced by external loading, causing eventually separation of the crack faces. This calculation is achieved by means of discrete physics of defects in contrast to the continuum physics principles used in [1] to arrive at partial differential equations for the hydrogen species.

A discrete model is also used for the calculation of the crack velocity, in contrast to the proposal in [2] which is based on a continuum interpretation. Here it is assumed that a crack velocity is attained by successive jumps of the crack tip of length less or equal to the length of the process zone when the mean concentration within this zone reaches a critical value. This requires a critical time interval for the hydrogen atoms to travel the distance of the process zone and accumulate within it until the mean conventration reaches its critical value. An estimate of this critical time is also given in terms of other known parameters of the model.

The results given here are rather suggestive and aim in illustrating some new appealing but still exploratory ideas in this difficult area of research. Comparisons with experiments and the results of the continuum models (e.g. [1]-[3]) will be made in a forthcoming article.

THE PROCESS ZONE AND THE STRUCTURE OF STRESS FIELD

In contrast to the elastic considerations in [1]-[3], we assume that the material is plastically hardened ahead of the crack tip. For sharp cracks, the extend of the plastic zone can be estimated (for example, on the basis of Irwin's model) as two orders of magnitude the length of the crack opening displacement. Moreover, for a plastic body (of the Ramberg-Osgood type) with work-hardening exponent n, the singular behavior of stresses and strains is described by [4]

$$[\sigma_{ij}, \sigma_e] = \sigma_o K_\sigma r^{-\frac{1}{n+1}} [\tilde{\sigma}_{ij}(\theta), \tilde{\sigma}_e(\theta)],$$

$$(1)$$

$$\varepsilon_{ij} = \alpha \varepsilon_o K_\varepsilon r^{-\frac{n}{n+1}} \tilde{\varepsilon}_{ij}(\theta),$$

where (r,θ) denote polar coordinates, $(\sigma_o, \varepsilon_o)$ reference values of stress and strain, $(\tilde{\sigma}_{ij}, \tilde{\sigma}_e, \tilde{\varepsilon}_{ij})$ specific functions of the angle θ, α a dimensionless constant, and the equivalent stress σ_e and plastic stress and strain intensity factors K_σ and K_ε are defined by the relations

$$\sigma_e = (\frac{3}{2} s_{ij} s_{ij})^{\frac{1}{2}}, \quad s_{ij} = \sigma_{ij} - \frac{1}{3}\sigma_{mm}\delta_{ij},$$

$$(2)$$

$$J = \alpha \sigma_o \varepsilon_o K_\sigma K_\varepsilon I_n = \alpha \sigma_o \varepsilon_o K_\sigma^{n+1} I_n = \alpha \sigma_o \varepsilon_o K_\varepsilon^{\frac{n+1}{n}} I_n,$$

with J in $(2)_2$ denoting the J-integral. The numerical factor I_n decreases monotonically from 6 (for $n \simeq 1$) to 4 (for $n \simeq 0$).

The singular stress distribution given by $(1)_1$ is induced by an external stress σ^∞ applied at infinity in a Mode-I situation. This solution does not consider the effect of the cohesive forces in the neighborhood of the crack tip which act in a direction opposite to the applied forces keeping the crack faces together. Barenblatt [6] was the first to introduce the concept of cohesive forces for brittle materials and this has led to the elimination of the elastic singularity. Barenblatt's concept has been extended by Panasyuk [7], Dugdale [8], Broberg [10], Smith [11], and others to incorporate a specific form of the cohesive forces including non-linear elastic and perfectly plastic behavior. These ideas have essentially led to postulating the existence of the "process zone", a small region ahead of the crack tip carrying a distribution of cohesive forces and controlling fracture. In the case of embrittlement, it is expected that hydrogen will modify the distribution of the cohesive forces and this was essentially assumed explicitly by Oriani et. al. [11] and implicitly by Aifantis et. al. [1]-[3]. A particular process zone was not recognized in [11] but an approximate treatment was utilized in [1]-[3] by identifying its length with the distance ahead of the crack tip where the concentration becomes maximum.

The concept of process zone and the distribution of cohesive forces within it, is considered on a more detailed mechanical basis below. It is not unreasonable to estimate the length r_p of the process zone (PZ) to be determined by the location of maximum stress ($\sigma_{22\,max}$) in the Rice-Johnson plastic solution [5] for blunted cracks as shown in figure 1. Not only the maximum stress occurs at this point, acting as hydrogen concentrator, but also an increase of the external load does not change the value of the maximum stress but merely its location, thus increasing the length of the process zone. This length turns out to be approximately twice the crack opening displacement and therefore the process zone is about two orders of magnitude smaller than the plastic zone. The distribution of cohesive forces $f(x_1)$ within the process zone $-r_p \leqslant x_1 \leqslant 0$ is assumed to be the linear of the form

$$f(x_1) = f_m + Mx_1, \quad M \equiv \tan \alpha = \frac{f_m}{r_{pc}}, \tag{3}$$

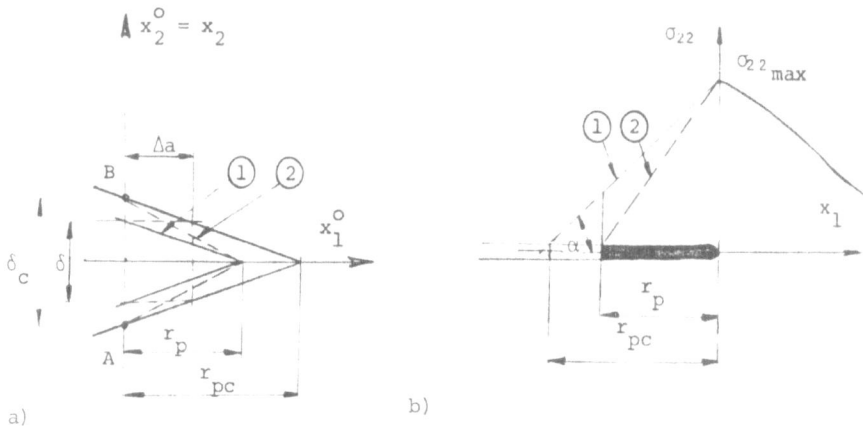

Fig. 1 a) Simplified profile of the PZ
        b) Scheme of the distribution
            of the PZ forces.
        Symbol 1 denotes the case without
                   hydrogen,
        Symbol 2 denotes the case with the
                   presence of the hydrogen.

with f_m denoting the maximum value of $f(x_1)$ at $x_1=0$ and r_{pc} denoting
the critical length of the process zone $f(-r_{pc})=0$ for critical external loading $(\sigma^\infty=\sigma_c^\infty)$. The cohesive force distribution along the crack
faces as given by (3) will also induce a singular stress distribution
determined by (1) and (2). On assigning the index (1) to the stress
distribution induced by the external loading and by (2) the stress
distribution induced by the cohesive forces we can eliminate the
plastic singularity at the tip of the process zone $(x_1= 0)$ for an
effective crack length $a+r_p$ by setting

$$J^{(1)}/I_n^{(1)} = J^{(2)}/I_n^{(2)}, \tag{4}$$

The value of $J^{(1)}$ can be estimated from an expression motivated by the Goldman-Hutchinson [12] analysis and in analogy to
practices in linear fracture mechanics. This expression reads [13]

$$J^{(1)} = \alpha\, \sigma_o \varepsilon_o\, (a+r_p)\, (\sigma^\infty/\sigma_o)^{n+1} \beta\, \hat{J}, \tag{5}$$

194

with the numerical factor $\hat{J} \equiv J(a/b \approx 0, n)$ given in [12] and β estimated [13] to vary between 1 and 1.25. The corresponding displacement field $u_2^{(1)}$ is given by

$$u_2^{(1)}(x_1) = \alpha \varepsilon_0 \left(\frac{a+r_p}{I_n^{(1)}}\right)^{\frac{n}{n+1}} \left(\frac{\sigma^\infty}{\sigma_0}\right)^{11} \beta \hat{J}(-x_1)^{\frac{1}{n+1}} , \qquad (6)$$

The value of $J^{(2)}$ is obtained by computing it along a contour shrinking to the boundary of the process zone. The result is

$$J^{(2)} = 2 \int_{-r_p}^{0} \frac{\partial u_2^{(2)}}{\partial x_1} f(x_1) dx_1 , \qquad (7)$$

To evaluate (7) we utilize (3) and an assumption similar to Smith's [10]

$$u_2^{(t)}(x_1) = u_2^{(t)}(-r_p) \left(-\frac{x_1}{r_p}\right)^{m+3/2} , \qquad (8)$$

together with (6) and an additivity hypothesis for the displacements of the form

$$u_2^{(t)} = u_2^{(1)} + u_2^{(2)} , \qquad (9)$$

Then, on utilizing (4) we arrive at the following consistency condition for m

$$m = \left\{ \frac{\left(\frac{n+1}{n+2}\right) B \left(\frac{\sigma^\infty}{\sigma_0}\right)^n r_p^{\frac{n+2}{n+1}} + \frac{r_{pc}}{2f_m} \left[A \frac{\sigma^\infty}{\sigma_0}^{n+1} -2f_m u_2^{(2)}(-r_p) \left(\frac{r_p}{r_{pc}} -1\right) \right]}{r_p u_2^{(t)}(-r_p)} \right\}^{-1} - \frac{5}{2} \qquad (10)$$

where

$$B \equiv \alpha \varepsilon_0 \left(\frac{a+r_p}{I_n^{(1)}}\right)^{\frac{n}{n+1}} \beta \hat{J} , \qquad A \equiv \frac{I_n^{(1)}}{I_n^{(2)}} \alpha \varepsilon_0 \sigma_0 (a+r_p) \beta \hat{J} ,$$

$$u_2^{(2)}(-r_p) = B\, r_p^{\frac{1}{n+1}} - \frac{1}{2}\, d_n(\frac{K^2}{\sigma_{ys} E}), \qquad r_p = (\frac{K}{\sigma_{ys}})^2 X_{max\sigma},$$

$$r_{pc} \equiv (\frac{K_{Ic}}{\sigma_{ys}})^2 X_{max\sigma}, \qquad K = 1.12\, \sigma\infty\, [\pi(a+r_p)]^{\frac{1}{2}}, \qquad (11)$$

The factor $X_{max\sigma}$ varies with the ratio σ_{ys}/E and for $\sigma_{ys} = 1300$ MPa it turns out that $X_{max\sigma} = \sigma_{ys}/E$. Moreover, we have

$$r_p \simeq 4\, u_2^{(t)}(-r_p), \quad u_2^{(t)} = \frac{1}{2}\, d_n(\frac{K^2}{\sigma_{ys} E}), \qquad (12)$$

The last relationship is an exact one following from $(11)_3$ with more details contained in [13].

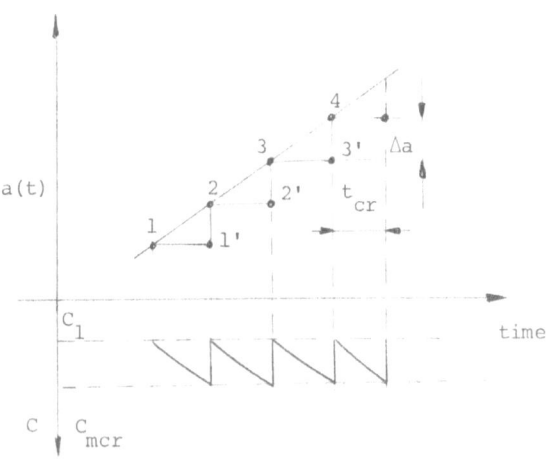

Fig. 2 Step-wise fashion crack tip motion, Δa-length of the each jump; t_{cr}-the time interval between two jumps following each other.

CRACK KINETICS

In contrast to the postulate adopted in [1]-[3] that the crack velocity is proportional to the value of hydrogen concentration at a critical distance (identified with the length of the process zone) ahead of the crack tip, a discrete analysis point of view is adopted here. We assume that the crack grows in a step-wise fashion as shown in Figure 2. If by Δa we denote the length of the crack jump and t_c the time necessary for such a jump to occur, we can write formally an expression for the mean velocity as

$$V_{mean} = \frac{\Delta a}{t_c} , \tag{13}$$

To calculate Δa, we refer to the simplified scheme of Figure 1 (with x_1^o measuring distances from the crack tip as opposed to x_1 which measures distances from the tip of the process zone) and note the relationship

$$2 u_2^{(t)} (x_1^o = \Delta a) \bigg|_{\sigma^\infty = \sigma_c^\infty} = \delta_o , \tag{14}$$

with $u_2^{(t)}$ given by (8) and δ_o denoting the crack tip opening displacement after the jump. Condition (14) determines the crack jump Δa. In the special case that the length of the crack jump equals the length of the process zone ($\Delta a = r_p$), δ_o is identified with the crack opening displacement in the absence of hydrogen for $\sigma^\infty < \sigma_c^\infty$ and can be determined from (12)$_2$ or equivalently from the relation

$$2 u_2^{(t)} (x_1^o = 0) \bigg|_{\sigma^\infty < \sigma_c^\infty} = \delta_o , \tag{15}$$

The critical time t_c is identified with the time required for the hydrogen to build up a critical mean concentration c_{mc} within the process zone. It is this mean critical concentration which stretches uniformly the lattice within the process zone and extends the crack faces from a distance δ_o to a distance δ_c where the critical event occurs (Figure 1). The extra displacement is provided by the dissolved hydrogen through the additivity assumption

$$\delta_c = \delta_o + \delta_H \ , \tag{16}$$

with δ_H denoting an average displacement measuring the extra
stretch of the lattice induced by the total amount of hydrogen
accumulated within the process zone. The lattice displacement
induced by a single hydrogen atom varies with the inverse of the
square of distance r

$$u^H = \frac{H}{r^2} \ , \tag{17}$$

where the constant H is a measure of the dipole moment characteristic
of the solute-matrix system. In the case that the full three-
dimensional displacement field $\underset{\sim}{u}^H$ is of interest, $\underset{\sim}{H}$ is a tensor
$\underset{\sim}{H} = (8\pi G)^{-1} \underset{\sim}{P}$ with G being the shear modulus and $\underset{\sim}{P}$ the dipole tensor
given by $P_{ij} = \lambda \delta_{ij}$ for spherical symmetry, $P_{ij} = (\lambda \delta_{ij} + \mu \delta_{1i} \delta_{1j})$
for tetragonal symmetry, and λ and μ being constants. For a uniform
distribution of hydrogen in the x_2 direction, we denote by $2r_o$ the
distance between two hydrogen atoms. Then

$$2r_o = \frac{1}{c(x_1, t)} \ , \tag{18}$$

with $c(x_1, t)$ denoting the concentration of hydrogen per unit length
at x_1. On defining an average concentration c_m by

$$c_m(t) = \frac{1}{r_p} \int_o^{r_p} c(x_1, t) \, dx_1, \tag{19}$$

we obtain for the average displacement u_m^H induced by the dissolved
hydrogen

$$u_m^H = 4Hc_m^2 \ , \tag{20}$$

The total displacement δ_H in the x_2 direction due to the presence
of hydrogen turns out to be

$$\delta_H = u_m^H c_m (\delta_o + \delta_c), \tag{21}$$

which in view of (20), and (16) yields

$$4H \ c_{mc}^{3} = \frac{1-\left(\delta_o/\delta_c\right)}{1+\left(\delta_o/\delta_c\right)} , \tag{22}$$

This last result together with the well-known relation $\delta_o/\delta_c = (\sigma\infty/\sigma\infty_c)$ also implied by $(12)_2$ gives an estimate for the mean critical concentration

$$c_{mc} = \left\{ \frac{1}{4H} \frac{[1-(\sigma\infty/\sigma\infty_c)^2]}{[1+(\sigma\infty/\sigma\infty_c)^2]} \right\}^{1/3} , \tag{23}$$

The velocity of a hydrogen atom travelling through the process zone is assumed to be constant due to the fact that the gradient of the cohesive forces is constant. This assumption is appropriate in a discrete situation where the main driving force is an existing stress gradient. In a continuum idealization both concentration and stress gradients are important in determining an appropriate expression for the flux [1]-[3]. While a continuum model to relate the mean critical concentration c_{mc} and the critical time t_c required to build up this concentration is possible, we list here a preliminary estimate obtained on a discrete basis. The details of this calculation will be presented elsewhere [13] where a comparison within a continuum model will also be made. In a simplified scheme where effects due to non-uniformities of concentration along the process zone are neglected, it can be shown that

$$t_c = \frac{c_{mc} \ \xi^{-1} - \overset{o}{c}(1-\xi)^{-1}}{\overset{o}{J}} + \frac{r_p}{v_H} , \tag{24}$$

with ξ denoting a trapping coefficient, $\overset{o}{c}$ and $\overset{o}{J}$ the concentration and flux of hydrogen at the entrance of the process zone, and v_H the velocity of hydrogen within it. The velocity v_H is proportional to M defined in (3) with the constant of proportionality determined by experiment or molecular considerations.

Numerical results illustrating the applicability of the model and comparisons with other theoretical attempts and experiments will be given in [13].

ACKNOWLEDGEMENTS

The support of the National Science Foundation and the Mechanics of Microstructures Program of the Michigan Technological University is gratefully acknowledged.

REFERENCES

1. Aifantis, E.C., Elementary Physicochemical Degradation Processes,
    in: Mechanics of Structured Media, Ed. A.P.S. Selvadurai, pp.
    301-317 (Amsterdam, Elsevier Sci. Pub. Co. 1981).

2. Unger, D.J. and E.C. Aifantis, On the Theory of Stress-Assisted
    Diffusion II, Acta Mechanica 47 (1983) 117-151.

3. Unger, D.J., Gerberich W.W. and E.C. Aifantis, Further Remarks
    on the Implications of Steady-State Stress-Assisted Diffusion on
    Environmental Cracking, Scripta Metallurgica 16 (1982) 1059-1064.

4. Hutchinson, J.W., Singular Behaviour at the End of a Tensile Crack
    in a Hardening Material, J. Mech. Phys. of Solids 16 (1968) 13-31

5. Rice, J.R. and M.A. Johnson, The Role of Large Crack Tip Geometry
    Changes in Plane Strain Fracture, Inelastic Behaviour of Solids
    Eds. M.F. Kanninen et. al. pp. 641-671 (New York, McGraw Hill
    1970).

6. Barenblatt, G.I., The Formation of Equilibrium Cracks During
    Brittle Fracture. General Ideas and Hypotheses, Axially
    Symmetric Cracks, PMM 23 (1959) 434-444.

7. Leonov, M.Ya. and V.V. Panasyuk, Development of Minute Cracks
    in Solids, Prikl. Mekhanica 5 (1959) 391-401.

8. Dugdale, D.S., Yielding of Steel Sheets Containing Slits,
    J. Mech. Phys. Sol., 8 (1960) 100 - 106.

9. Broberg, K.G., Crack-Growth Criteria and Non-linear Fracture
    Mechanics, J. Mech. Phys. Solids, 19 (1971) 407-418.

10. Smith, E., The Structure in the Vicinity of a Crack Tip,
    Engineering Fracture Mechanics, 7 (1975) 285-289.

11. Oriani, R.A. and P.H. Josephic, Equilibrium Aspects of Hydrogen
    Induced Cracking of Steels, Acta. Metall. 22 (1974) 1065-1074.

12. Goldman, N.L. and J.W. Hutchinson, Fully Plastic Crack
    Problem: The Center-Cracked Strip Under Plane Strain, Int.
    J. Solids Structures 11 (1975) 575-591.

13. Neimitz, A. and E.C. Aifantis, Forthcoming.

APPENDIX

In this appendix we list some results that will be reported in a forthcoming publication which is a natural extension of the present paper [13].

It turns out that for most materials the exponent $m + 3/2$ in (8) should be replaced by m^* ($m^* > 1$). Moreover, to maintain consistency with the smooth clousure condition at $x_1 = 0$, the cohesive zone stresses at the crack tip σ_{22*} ($x_1 = r_p$), for the critical moment (Fig. 1.6) should be greater than zero. The magnitudes of these stresses depend on the value of m^* and other material properties (e.g. $\sigma_0/E, n$).

The exponent m^* can be calculated either from (10) or from the formula:

$$m^* = \frac{\zeta d_n - 1}{1 - \eta d_n} \; ; \quad \sigma_{22m} = \zeta \sigma_0 \; ; \quad \sigma_{22*} = \eta \sigma_0 \qquad (A.1)$$

where ($\zeta = 5.2$ for $n = 5$ and $\sigma_0/E = 0,0025$) or

($\zeta = 3.5$ for $n = 10$ and $\sigma_0/E = 0,0025$) and d_n

is determined from the well known equation:

$\delta = 2d_n (J/\sigma_0)$. Since $m^* > 1$, η can be calculated from (A.1).

Thus, for $\zeta = 5.2$ and $m^* = 1,2$ it follows that $\eta = 1.076$, while for $\zeta = 3.5$ and $m^* = 1.66$ it follows that $\eta = 1.0$. The more complete Table will be provided in [13].

Equation (A.1) can be derived from (10) for critical condition (at the onset of jump) if the crack becomes semi-infinite or directly from the formula

$$J = J_{Ic} = \int_{\Gamma end} f(x_1) \; \frac{du_2}{dx_1} \; dx_1 \qquad (A.2)$$

by using (3) and (8) as they are now modified by the arguments given in the beginning of the Appendix.

The present model was utilized to predict the graphs of K_{TH} v.s. external hydrogen pressure in several materials. Also the crack tip velocity was found as a function of the above parameters. Details of these calculations will also be included in [13].

A REVIEW OF FATIGUE CRACK INITIATION

J.W. Provan   Z.H. Zhai
Associate Professor   Graduate Student

Mechanical Engineering Department
McGill University
817 Sherbrooke St. West
Montreal, Que., Canada, H3A 2K6

ABSTRACT

This paper reviews the fundamental aspects of fatigue crack initiation in lightly alloyed polycrystalline metals. The introduction presents the motivations behind this study along with some general remarks. The second section discusses the mechanisms of fatigue crack initiation that include: i) the cyclic hardening and saturation of materials at early stages of fatigue loading; ii) the low strain, high cycle fatigue which includes a discussion of the behavior and structure of the persistant slip bands (PSBs), coupled with possible reasons for the formation of PSBs and extrusions/intrusions at boundaries of the PSBs; iii) the high strain, low cycle fatigue; iv) the nucleation of microcracks in polycrystalline materials for both the low strain and high strain situations; and v) theories of irreversibility of dislocation slip. The final section draws some conclusions concerning this study.

INTRODUCTION

Fatigue fracture causes not only injuries but also financial losses in many industries such as the aircraft, nuclear and pulp and paper. For example, an investigation showed that in aircraft, 25 to 50 percent of all structural failures are the result of fatigue [1].

Crack initiation plays an important role in the whole fatigue life of either a component in a structure or a specimen in a laboratory. Though some researchers consider that cracks initiate at very early stages in a fatigue situation, from 5 to

15 percent for both pure materials and alloys [2,3], others consider that crack initiation and coalescence of microcracks lead to a macroscopic crack over a wide range of roughly the first 90% of the total fatigue life [4-6]. Although this appears contradictory it is not since it depends on the definition of the initiation of a crack. For example, Forsyth [3] considered that those cracks forming early are only microcracks and that it takes a long time for them to become macroscopic in dimension. If the definition of crack initiation is accepted as the formation of a macroscopic crack instead of a microscopic crack, then some of the confusion concerning crack initiation would be clarified. From this, the importance of crack initiation is quite clear.

It is well known that the experimental data of fatigue tests contains a large degree of scatter, therefore, a statistical method is necessary in fatigue analysis. The Provan reliability law was proposed to analytically describe this phenomenon [7]. This law was derived on the basis of the micromechanics philosophy of fatigue crack initiation and a Markov linear birth stochastic crack propagation process. The failure density of a component is given by this law as:

$$p_{N_p}(i) = \frac{\mu_{a_0}}{\sqrt{2\pi}V_{ac}} \exp\left\{\lambda i - \frac{(\mu_{a_0}\exp[\lambda i] - \mu_{a_f})^2}{2V_{ac}}\right\} \qquad (1)$$

where:

V_{ac} = variance of crack penetration
μ_{ao} = mean of crack initiation depth
λ = growth rate transition intensity
and μ_{af} = mean of critical crack depth.

Experiments were performed to check on the validity of this law [8,9]. By curve fitting the Provan reliability law to the experimental points, the transition intensity parameter in Eqn. (1) was obtained as:

$$\lambda_{emp} = 0.20 \times 10^{-3}/\text{cycle}; \quad \Delta\varepsilon/2 = 0.003, \qquad (2)$$

while by an experimental fractographic procedure described in [10], the value of λ was evaluated as:

$$\lambda_{exp} = 1.89 \times 10^{-3}/\text{cycle}; \quad \Delta\varepsilon/2 = 0.003. \qquad (3)$$

Comparing (2) and (3), the discrepency is one order of magnitude. The reason for this discrepency is the underestimation of the role of crack initiation [6]. One of the motivations of this paper is to study the mechanisms of fatigue crack initiation so that the Provan reliability law can be refined.

Furthermore, while it is relatively easy to determine the mechanical properties of a material such as the elastic modulus, yield strength and ultimate strength, the experimental scatter from other tests is large enough to warrant in-depth investigations. This is especially true in the situation where fatigue damage is being monitored. From the onset of cyclic loading, fatigue damage starts by changing the microstructure of the material. Much effort has been extended on the mechanisms of fatigue and there are now good reviews on the subject of fatigue mechanisms [2,11,12]. Although much progress has been made, a general observation is that due to the complexity of the fatigue damage mechanism, this area of investigation is still at a qualitative rather than at a quantitative stage.

Fatigue crack initiation is, then, one of the major aspects of all fatigue studies. This paper reviews some of the outstanding work previously done in this field with the emphasis being placed on a systematization of fatigue crack initiation theories.

MECHANISMS AND THEORIES OF FATIGUE CRACK INITIATION

Strain Hardening and the Cyclic Stress Strain Curve

For many single crystals a rapid hardening occurs at the early stages of the fatigue process [13-18]. The hardening rate then slows down until it becomes zero when the material is said to have reached its **saturation** stage [16]. Furthermore, Hancock and Grosskreutz [15] found that the dislocation structure in the earliest stages of rapid hardening were analogous to unidirectional dislocation structures. They also found the dislocation bundles in dipole and multipole configurations contributing to rapid hardening by providing effective barriers to continued dislocation motion. Typical hysteresis loops associated with this hardening for an annealed copper single crystal are shown in [19], where it is clearly shown that the hardening slows down as the cycles increase.

For different strain levels, one corresponding saturation stress exists. By connecting these strains to the saturation stresses, a cyclic stress strain curve can be obtained as discussed by Mughrabi [16]. There are three regions in the curve which is very similar to a creep curve. Regions A and B correspond to low strain, high cycle fatigue and region C corresponds to high strain, low cycle fatigue. Region B is known as the **plateau** in which the stress is almost strain independent.

It is worthwhile to note that strain hardening is the precursor of fatigue crack initiation. Indeed, Alden and Backofen [20] found that crack nucleation can be prevented by periodically annealing the specimen thereby keeping it in a work hardenable state. Furthermore, for the strain levels within the plateau region, Watt et al [21], observed that the micromechanical structure of the material changed so that the **Persistent Slip Bands (PSBs)** formed at the end of the hardening process. The PSBs are so important to an understanding of the fatigue mechanism that they deserve a detailed discussion in the next section.

Low Strain Amplitude, High Cycle Fatigue Crack Initiation

Persistent Slip Bands (PSBs). In copper and other face-centered-cubic (fcc) and in some body-centered-cubic (bcc) structured materials, PSBs are often observed in low strain amplitude fatigue. The name PSBs comes from the "persistence" of some of the slip bands even after a thin layer of the surface has been removed by electropolishing during fatigue tests. The formation and performance of PSBs have been widely studied due to the fact that for low strain amplitude fatigue, cracks usually initiate at locations where the PSBs interact with surface boundaries, such as free surfaces or grain boundaries. When PSBs intersect with a free surface, the microcracks nucleate **transgranularly,** while in the grain boundary case **intergranular** cracking has been observed in both fine and coarse grained materials [22,23]. Typical pictures of PSBs are given in [24-27] where the ladder like structure of PSBs comprising of high dislocation density walls and low dislocation density matrix material are illustrated.

The screw dislocation density between the walls of PSBs is as high as $\sim 10^{13}$ m^{-2} whereas the density between the veins of the bulk material is $\sim 10^{12}$ m^{-2}. Moreover, the dislocation densities in the veins and the walls are almost the same, $\sim 10^{15}$ m^{-2} [28,29]. Therefore, since the volume fraction of hard walls in the PSBs is only 0.1 compared to the 0.5 volume fraction of hard veins in the bulk matrix, the PSBs are much softer than the bulk material and hence the strain is localized in the PSBs. Kuhlmann-Wilsdorf and Laird [27] have proposed a model which describes the dislocation motion inside PSBs. Briefly, when a shear stress acts on a plane containing PSBs, the edge dislocations on one side of a PSB wall are "spun out" by many similarly signed screw dislocations to the nearby tilt wall containing edge dislocations of opposite sign, thereby allowing the large strains observed in the neighbourhood of PSBs. For a stress reversal, the direction of the "spin" is reversed.

From Neumann's experimental results [30,31] which showed bursts of strain induced by stress increase, Kuhlmann-Wilsdorf and Laird [27] proposed the following PSB formation theory: In the early stages of the cyclic loading process, loop patches form to block the dislocations. As the stress slowly increases to a certain level, these loop patches are destroyed accompanied by the occurrence of strain bursts. When the cyclic loading continues, the loop patches are established again with a finer hence stronger structure. At a still higher stress level, the new loop patches will be destroyed again. This build/ destroy procedure repeats many times until the strain becomes so high that no loop patches can stop the movement of the dislocations. Later, the same authors [32] determined that the transformation of the loop patches into PSBs need not be catastrophic but can take place gradually and continuously.

As for how the PSBs are formed from the loop patchs or vein structure, Kuhlmann-Wilsdorf and Laird [27,32] supported Winter's [33] proposal that PSBs are formed from the inside of the veins or loop patches. Mughrabi et al [25], on the other hand argued that PSBs are formed from outside of the veins.

Another question concerns the critical value at which PSBs are formed. Mughrabi et al [25] believes that the veins cannot accomodate plastic strain in excess of $\sim 10^{-4}$. If the plastic strain becomes larger than this value, the structure will undergo a rearrangement and PSBs will form since, as discussed above, the PSBs are more capable of carrying large amounts of strain. Contradictory to this, Kuhlmann-Wilsdorf and Laird [32] found that for copper single crystals, PSBs begin to form when the applied resolved shear stress reached about 30 MPa on the primary slip system. They conjectured that this value depends little, if at all, on the strain amplitude within the PSB and the rate of cycling and, therefore, this critical stress appears to be triggered at a critical loop density. As can be seen, a considerable amount of controversy still surrounds this topic and only further experimentation will resolve this issue.

Extrusions/Intrusions at the Surface of Fatigued Specimens. For low strain amplitude fatigue, the formation of extrusions/ intrusions is an integral step towards crack initiation. Once extrusions/intrusions are formed, they act as stress raisers at the surface and microcracks tend to occur at these sites. By repolishing the surface to eliminate extrusions/intrusions the life of a specimen can be considerably extended [34].

Many extrusion/intrusion models have been proposed. In low strain amplitude fatigue, extrusions/intrusions are found mainly to occur at the boundary of PSBs [35-37]. Two extrusion/ intrusion models are of particular interest since both are based on the movement of the dislocations inside PSBs.

Kuhlmann-Wilsdorf and Laird [27], following their model of the structure of the PSBs and the theory of dislocation movement within them, deduced that extrusions/intrusions are formed by gliding of the edge dislocations with the help of the motion of screw dislocations out of the PSB boundary or into the boundary, respectively. Essmann et al [37], alternatively interpreted extrusions/intrusions as a combination of gliding and annihilation of dislocations within the slip bands.

In [23], Mughrabi et al also observed that extrusions formed early in fatigue tests. They found that for pure copper, extrusions continue to grow but at a lower rate as the fatigue cycles increased. Another noticable point in their paper is that only extrusions were observed with almost no intrusions on the surface of fatigued specimens.

There are still some other problems in extrusions/ intrusions theory. For example, although the above two theories seem to be applicable for low strain amplitude fatigue, they can not explain the occurrence of extrusions/intrusions for high strain fatigue [38] where PSBs are not usually observed [39].

High Strain Amplitude Fatigue Crack Initiation

In high strain amplitude fatigue, the specimen usually possesses a cell structure and the deformation, as a result, tends to be homogeneous, which differs from low strain fatigue where the strain is localized in PSBs. Although some extrusions are also observed in high strain fatigue, they are not severe enough to cause transgranular cracking, instead, notches or steps at boundaries formed by the slip irreversibility create stress raisers and effective sites for crack initiation [40-42]. As a consequence, fracture is most often found at grain boundaries; in other words, fracture occurs intergranularly in the high strain fatigue situation.

The Microcrack Nucleation Process

For low strain amplitude fatigue, a microcrack nucleates along an active slip plane, generally a PSB [11]. As long as the crack develops along the PSBs, which are usually at an angle of 45° to the applied stress direction, this crack is considered by the current authors to be nucleating. (In some other papers, this is called Stage I propagation). The nucleation process terminates when the crack becomes large enough to change directions and propagate normally to the applied stress. The corresponding mechanism of crack nucleation along a PSB is discussed by Laird [12]. In this latter model, the total volume remains unchanged since the extrusions are exactly balanced by the intrusions that form during the process.

For high strain amplitude fatigue, Kim and Laird [41,42] define the "Stage I propagation" as the stage of crack propagation before plastic blunting at the crack tip occurs, since the crack nucleation and growth for high strain fatigue are not on crystallographic planes. At a high strain amplitude level, **persistent grain boundaries** are observed instead of PSBs [43]. Cracks are expected to grow along these boundaries both on the surface and into the bulk simultaneously. Newmann's model [31] can be used to explain the mechanisms of high strain fatigue crack nucleation.

Finally, for many lightly alloyed metals of engineering importance, inclusions are regarded as stress and strain concentrators. Voids form around these inclusions [44], and the crack nucleation process for these second phase materials is actually a process of void coalescence [45].

Reversibility and Irreversibility of Cyclic Slip

Finney and Laird [46] studied the localization of plastic strain in copper single crystals strain cycled into saturation and found that beyond a certain level of strain there was an overall reversibility and some individual slip step irreversibility. What induces this irreversibility is not very clear. Margolin et al [47] suggested that in the irreversible slip process, cross-slip is a control factor. Some other authors [23,26,48], believe that the mutual annihilation of unlike dislocations constitutes a major source of cycle irreversibility. This annililation persists in cyclic saturation since a dynamic balance exists between dislocation multiplication and annihilation in steady-state cyclic saturation.

A parameter, β, associated with the annihilation distance, Burger's vector, the number of dislocations of the same sign and the shear strain amplitude has been proposed to determine whether the motion of the dislocation is reversible or not [23,48,49]. For $\beta < 1$, the dislocations move reversibly whereas for $\beta > 1$, slip becomes irreversible.

Following some experimental investigations, Mughrabi et al [23] draw the conclusion that in single crystals, surface roughening of the extruded material is caused mainly by the random irreversible slip, both the surface roughening and random irreversible slip being responsible for the fatigue damage.

CONCLUDING REMARKS

For low strain amplitude fatigue, cyclic hardening starts fatigue damage by changing the microstructure of a material from a vein structure to a PSB structure. The PSB has a ladder-like structure with highly concentrated dipole edge dislocations in the walls and carries highly localized strain, probably through the "spinning" of dipolar edge dislocation walls. Extrusions/intrusions are formed at the surface boundaries of PSBs by sliding and annihilation of dislocations and act as stress raisers to initiate microcracks. Once these cracks are initiated, they nucleate along slip planes to form macrocracks thus completing the crack initiation process.

208

For high strain amplitude , less is understood since less has been done. There are some features similar to low strain amplitude fatigue, such as the hardening and saturation phenomena, but the microstructures are different. A cell structure for high strain replaces the PSB structure for low strain fatigue. Cracks usually start at grain boundaries in the high strain fatigue situation.

Although what has been described above appears clear and simple, it is merely a skeleton with many details to be clarified. On some of the better researched details, controversy still exist and, therefore, more work is needed to give a thorough understanding of the mechanisms of fatigue crack initiation.

REFERENCES

[1] Cooper, T.D. and Kelto, C.A., "Fatigue in Machines and Structures - Aircraft", in **Fatigue and Microstructure, Material Science Seminar**, St. Louis, 1978, American Society for Metals, (1979), 29-56.

[2] Laird, C. and Duquette, D.J., "Mechanisms of Fatigue Crack Nucleation", in **Corrosion Fatigue: Chemistry, Mechanics and Microstructure**, eds., A.J. McEvily and R.W. Staehle, NACE, Houston, (1972), 88.

[3] Forsyth, P.J.E., "The Physical Basis of Metal Fatigue", Blackie & Sons Ltd., London, (1969).

[4] Buck, O., Morris, W.L. and James, M.R., "Remaining Life Prediction in the Microcrack Initiation Regime", in **Fracture and Failure: Analyses, Mechanisms and Applications**, eds., P.P. Tung et al, American Society for Metals, (1981), 55-64.

[5] Field, J.L., Behnaz, F. and Pangborn, R.N., "Characterrization of Microplasticity Developed During Fatigue", in **Fatigue Mechanisms: Advances in Quantitative Measurement of Physical Damage**, ASTM STP 811, American Society for Testing and Materials, (1983), 71-94.

[6] Provan, J.W., "The Micromechanics of Fatigue Crack Initiation", to be published in the Proceedings of the 10th Canadian Fracture Conference on "Modelling Problems in Crack Tip Mechanics", University of Waterloo.

[7] Provan, J.W., "A Fatigue Reliability Distribution Based on Probabilistic Micromechanics", in **Defects and Fracture**, eds., G.C. Sih and H. Zorski, Martinus Nijhoff Publishers, The Hague, (1982), 63-69.

[8] Provan, J.W. and Theriault, Y., "An Experimental Investigation of Fatigue Reliability Laws", in **Defects, Fracture and Fatigue,** eds., G.C. Sih and J.W. Provan, Martinus Nijhoff Publishers, The Hague, (1983), 423-432.

[9] Theriault, Y., "An Experimental Investigation of Fatigue Reliability Laws", Master's Thesis, McGill University, (1983).

[10] Provan. J.W., "The Micromechanics Approach to the Fatigue Failure of Polycrystalline Metals", Chapter 6 of **Cavities and Cracks in Creep and Fatigue,** ed., J. Gittus, Applied Science Publishers, London, (1981), 197-242.

[11] Thompson, N. and Wadsworth, N.J., "Metal Fatigue", Advances in Physics, $\underline{7}$, (1958), 72.

[12] Laird, C., "Mechanisms and Theories of Fatigue", in **Fatigue and Microstructure, Material Science Seminar,** St. Louis, 1978, American Society for Metals, (1979), 149-203.

[13] Kemsley, D.S. and Paterson, M.S., "The Influence of Strain Amplitude on the Work Hardening of Copper Crystals in Alternating Tension and Compression", Acta Met., $\underline{8}$, (1960), 453.

[14] Wadsworth, N.J., "Work Hardening of Copper Crystals under Cyclic Straining", Acta Met., $\underline{11}$, (1963), 663.

[15] Hancock, J.R. and Grosskreutz, J.C., "Mechanisms of Fatigue Hardening in Copper Single Crystals", Acta Met., $\underline{15}$, (1967), 1275.

[16] Mughrabi, H., "The Cyclic Hardening and Saturation Behavior of Copper Single Crystals", Mater. Sci. Eng., $\underline{33}$, (1978), 207.

[17] Feltner, C.E., "A Debris Mechanisms of Cyclic Strain Hardening for F.C.C. Metals", Phil. Mag., $\underline{12}$, (1965), 1229.

[18] Basinski, S.J., Basinski, Z.S. and Howie, A., "Early Stages of Fatigue in Copper Single Crystals", Phil. Mag., $\underline{19}$, (1969), 899.

[19] Finney, J.M., "Strain Localization in Cyclic Deformation", Ph.D. Thesis, Univ. of Pennsylvania, (1974).

[20] Alden, T.H. and Backofen, W.A., "The Formation of Fatigue Cracks in Aluminum Single Crystals", Acta Met., $\underline{9}$, (1961), 352.

[21] Watt, D.F., Embury, J.D. and Ham, R.K., "The Relation Between Surface and Interior Structures in Low-Amplitude Fatigue", Phil. Mag., 17, (1968), 199.

[22] Mughrabi, H., "A Model of High-Cycle Fatigue-Crack Initiation at Grain Boundaries by Persistent Slip Bands", in **Defects, Fracture and Fatigue**, eds., G.C. Sih and J.W. Provan, Martinus Nijhoff Publishers, The Hague, (1983), 139-146.

[23] Mughrabi, H., Wang, R., Differt, K. and Essmann, U., "Fatigue Crack Initiation by Cyclic Slip Irreversibilities in High-cycle Fatigue", in **Fatigue Mechanisms: Advances in Quantitative Measurement of Physical Damage**, ASTM STP 811, American Society for Testing and Materials, (1983), 5-45.

[24] Pohl, K., Mayr, P. and Macherauch, E., "Shape and Structure of Persistent Slip Bands in Iron Carbon Alloys", in **Defects, Fracture and Fatigue**, eds., G.C. Sih and J.W. Provan, Martinus Nijhoff Publishers, The Hague, (1983), 147-159.

[25] Mughrabi, H., Ackermann, F. and Herz, K., "Persistent Slip Bands in Fatigued Face-Centered and Body-Centered Cubic Metals", in **Fatigue Mechanisms**, ed., J.T. Fong, ASTM STP 675, American Society for Testing and Materials, (1979), 69-105.

[26] Mughrabi, H. and Wang, R., "Cyclic Strain Localization and Fatigue Crack Initiation in Persistent Slip Bands in Face-Centered Cubic Metals and Single-phase Alloys", in **Defects and Fracture**, eds., G.C. Sih and H. Zorski, Martinus Nijhoff Publishers, The Hague, (1982), 15-28.

[27] Kuhlmann-Wilsdorf, D. and Laird, C., "Dislocation Behavior in Fatigue", Mat. Sci. Eng., 27, (1977), 137-156.

[28] Woods, P.J., "Low-Amplitude Fatigue of Copper and Copper-5% Aluminum Single Crystals", Phil. Mag., 28, (1973), 155.

[29] Antonopoulos, J.G. and Winter, A.T., "Weak-Beam Study of Dislocation Structures in Fatigued Copper", Phil. Mag. 33, (1976), 87-95.

[30] Neumann, P., "Strain Bursts and Coarse Slip During Cyclic Deformation", Z. Metallkd., 59, (1968), 927.

[31] Neumann, P., "Coarse Slip Model of Fatigue", Acta Metall., 17, (1969), 1219.

[32] Kuhlmann-Wilsdorf, D. and Laird, C., "Dislocation Behavior in Fatigue V: Breakdown of Loop Patches and Formation of Persistent Slip Bands and of Dislocation Cells", Mat. Sci. Eng., 46, (1980), 209.

[33] Winter, A.T., "Nucleation of Persistent Slip Bands in Cyclically Deformed Copper Crystals", Phil. Mag., 37, (1978), 457.

[34] Thompson, N., Wadsworth, N. and Louat, N., "The Origin of Fatigue Fracture in Copper", Phil. Mag. 1, (1956), 113.

[35] Forsyth, P.J.E., "Extrusion of Material from Slip Bands at the Surface of Fatigued Crystals of an Aluminum-Copper Alloy", Nature, 171, (1953), 172.

[36] Steiner, D. and Gerold, V., "Fatigue Softening in Precipitation Hardened Copper-Cobalt Single Crystals", in **Defects, Fracture and Fatigue**, eds., G.C. Sih and J.W. Provan, Martinus Nijhoff Publishers, The Hague, (1983), 183-194.

[37] Essmann, U., Goesele, U. and Mughrabi, H., "A Model of Extrusion and Intrusion in Fatigued Metals, Part 1", Phil. Mag. A, 44, (1981), 405.

[38] Laird, C. and Smith, G.C., "Initial Stages of Damage in High Stress Fatigue in some Pure Metals", Phil., Mag. 8, (1963), 1945.

[39] Kemsley, D.S., "The Nature of Persistent Slip Bands in Fatigued Copper", Phil. Mag., 2, (1957), 131.

[40] Kemsley, D.S., "Crack Paths in Fatigued Copper", J. Inst. Met., 85, (1956), 420.

[41] Kim, W.H. and Laird, C., "Crack nucleation and Stage I Propagation in High Strain Fatigue, I", Acta Met., 26, (1978), 777-787.

[42] Kim, W.H. and Laird, C., "Crack nucleation and Stage I Propagation in High Strain Fatigue, II", Acta Met., 26, (1978), 789-799.

[43] Smith, G.C., "The Initial Fatigue Crack", Proc. R. Soc. A 242, (1957), 189.

[44] Mitchell, M.R., "A Unified Predictive Technique for the Fatigue Resistance of Cast Ferrous-Based Metals and High Hardness Wrought Steels", SAE/SP-79/448, Paper No. 790890, (1979).

[45] Lynch, S.P., "Mechanisms of Fatigue and Environmentally Assisted Fatigue", in **Fatigue Mechanisms**, ed., J.T. Fong, ASTM STP 675, American Society for Testing and Materials, (1979), 174-213.

[46] Finney, J.M. and Laird, C., "Strain Localization in Cyclic Deformation of Copper Single Crystals", Phil. Mag., 31, 8th Series, (1975), 339-366.

[47] Margolin, H., Mahajan, Y. and Saleh, Y., "Grain Boundaries, Stress Gradients and Fatigue Crack Initiation", Scripta Met., 10, (1976), 1115.

[48] Essmann, U. and Mughrabi, H., "Annihilation of Dislocations during Tensile and Cyclic Deformation and Limits of Dislocation Densities", Phil. Mag. A, 40, (1979), 731-756.

[49] Essmann, U., "Irreversibility of Cyclic Slip in Persistent Slip Bands of Fatigued Pure F.C.C. Metals", Phil. Mag. A, 45, (1982), 171-190.

CREEP CRACK GROWTH IN DUCTILE MATERIALS

V.M. Radhakrishnan.

Metallurgy Department,
Indian Institute of Technology,
Madras-600036, India.

ABSTRACT

Many parameters such as stress intensity factor, net section stress, reference stress and J* have been proposed to describe the crack growth in creep. However, large and systematic scatter has been observed in their description of the experiemntal data. Derivation of a parameter R based on the first principle is described and the discrepancy between the experimental data and the other parameters, specially the J* is discussed.

KEY WORDS

Creep crack, energy approach, J* integral for creep, elastic plastic fracture mechanics.

INTRODUCTION

Crack growth in creep has interested many roooorohoro, ao several in-service failures in power generating plants have been found to be due to the growth of creep cracks. Several parameters such as K, net section stress, reference stress, COD etc., have been tried to describe the crack growth rate, but only with limited success. A typical relation between da/dt and K as represented by the scatter bands, is shown in Fig.1, (data from reference 1-4) which clearly shows that there is no unique relationship between the two variables. The path independent line integral J has been extended to creep crack growth (5-7) and for a given displacement rate $\dot{\Delta}$, it is given as

$$J* = -d\dot{U}/B \, da \, \big|_{\dot{\Delta}_1} \tag{1}$$

where $\dot{U}$ is the energy associated with load P-deflection rate curve and B the thickness of the specimen. Many attempts have been made to evaluate the energy rate line integral $J*$. Kotera-zawa and Mori (6) give the relation for CC and CT type of specimen as

$$J* = \frac{n-1}{n+1} \, \sigma_{net} \dot{\Delta} \propto \frac{P\dot{\Delta}}{B(W-a)} \tag{2}$$

where W is the width of the specimen, a the crack length and n the creep exponent.

Nikbin et al (2) evaluated the relation for TDCB specimens in the form

$$J* = \frac{\eta}{B(n+1)} \, \frac{P\dot{\Delta}}{a} \tag{3}$$

where η and n are constants. Analysis by Hutchinson et al (7) leads to an expression for CC specimen in the form

$$J* = A \, \frac{P^{n+1} a}{(W-a)^n} \, g(a/W) \tag{4}$$

$$= A_1 P \dot{\Delta} a$$

where A is a constant and g a function of (a/W). Creep crack growth rate da/dt has been correlated with $J*$. A few typical published results under load controlled conditions are shown in Figs. 2,3, and 4. The raw data are from Ref(4,8,9) and the full lines in the figures are due to the present author. It can be seen that in all the cases there is systematic scatter and the relation da/dt vs $J*$ is load dependent. For a given load the relation is non-linear on the log-log plot. In general there is large scatter and in the typical example of 316 type stainless steel, the scatter is by a factor of 50.

PARAMETER DERIVATION

Fig.5 shows the data reduction procedure from the first principle. The relation between $\dot{U}$ and a can be obtained in three possible conditions, i.e., (a) a convex curve (b) a strai-ght line or (c) a concave curve for a given $\dot{\Delta}$. Under condition (a) d$\dot{U}$/da at higher P will be smaller than at lower P and vise versa in condition (c). The corresponding relation between $\dot{\Delta}$ and $\dot{a}$ is shown in the figure at the bottom. In condition (a) $\dot{a}$ will be smaller at higher loads and opposite is the case in condition (c). The relation between a and can be given by

$$\dot{a} \propto \dot{\Delta}/(P) \qquad \text{for condition (a)} \tag{5a}$$

$$\dot{a} \propto (\dot{\Delta}) \quad \text{for condition (b)} \quad (5b)$$

$$\dot{\Delta}(P) \quad \text{for condition (c)} \quad (5c)$$

The brackets indicate functions. The possibility (b) has been observed for creep brittle materials (10,11) where there is very little plastic deformation in the crack front region and the crack growth rate is a function of COD rate or load point deflection rate only. The relation between $\dot{U}$ and $\underline{a}$ for 316 type stainless steel, the raw data taken from Ref(12) is shown in Fig.6., which supports the possibility (a) in ductile materials. The relation between $\dot{\Delta}$ and $\dot{a}$ is shown in Figs. 7 and 8 for 316 stainless steel and 6061 Al alloy, the data taken from Ref (12,13). The load dependence of these relations clearly indicate the possibility (a) in the case of ductile materials.

DISCUSSION

Consider a body with a crack subjected to creep deformation. The deformation is localised near the net section and the crack is continuously growing, as it happens in a bending type of loading. The isochronous load P - deflection Δ relation, shown schematically in Fig.9, can be given by

$$\Delta = C_{(t)} (P/P_o)^{\alpha} \quad (6)$$

where the compliance C is time dependent. For constant load condition the deflection rate can be given as

$$d\Delta/dt = d\Delta/dC \cdot dC/dt = \dot{C} (P/P_o)^{\alpha} \quad (7)$$

With time both the crack length and the compliance increase. The change in compliance dC with change in crack length da at larger values of $\underline{a}$ will be more than at smaller values of $\underline{a}$. So if the compliance is taken as (Fig.10)

$$C \propto (a)^{\gamma} \quad (8)$$

where the exponent γ is a constant, we get

$$\frac{dC}{dt} = \frac{dC}{da} \frac{da}{dt} = \gamma (a/a_o)^{\gamma - 1} \dot{a} \quad (9)$$

where a_o is a constant. Combining (7) and (9) and rearranging we get

$$\dot{a} \propto \dot{\Delta} / (P/P_o)^{\alpha} (a/a_o)^{\gamma - 1} = R \quad (10)$$

In the case of bending type of loading, the parameter R becomes

$$R = \dot{\theta} / (M/M_o)^{\alpha} (a/a_o)^{\gamma - 1} \quad (11)$$

where M is the bending moment per unit thickness and $\dot{\theta}$ is the deflection rate.

The important assumption made in the above analysis is that the compliance C which increases with time can be represented by another variable $\underline{a}$ which also increases with time, as given by equation (8). The reasonableness of this assumption can be explained as follows. Considering equation (6) the compliance is equal to the load point deflection at load $P = P_o$, where P_o is a reference load which can be taken to be small enough, causing negligible creep deformation. So we have

$$\Delta = C_{(t)} \big|_{P=P_o} \qquad (12)$$

In such a case, any increase in the load point deflection Δ will be caused only due to increase in crack length $\underline{a}$ and it can be given by

$$\Delta \propto (a)^{\gamma} \qquad (13)$$

So it can be taken that the compliance $C_{(t)}$ is also dependent on the crack length $\underline{a}$ through a similar function so that

$$C_{(t)} \propto (a)^{\gamma} \qquad (8)$$

The relation between the crack growth rate and the para-meter R is shown in Fig. 11 for 6061 Al alloy for two types of specimen geometries, namely, CC and CT specimens (14). The exponent γ for the material investigated is taken as unity under the test conditions. The description appears to be very good for the specimens subjected to bending type of loading and the creep deformation is confined to the crack region.

Combining equations (2) and (10) we get

$$\dot{a} \propto J* \frac{(W-a)}{P^{\alpha+1} \, a^{\gamma-1}} \propto J* \beta \qquad (14)$$

It can be seen that the factor β will decrease as $\underline{a}$ increases for a given load P. So if β is not considered, then $\dot{a}$ vs J* relation will have a larger value on the J* axis for a given crack and this will result in a drooping or flattening curve as the crack grows. Such a trend has been observed in Figs.2, 3 and 4. β is also load dependent - it decreases with increasing load. It can be seen in the figures that the curves for higher loads get shifted to the right. If β is also included as indicated in equation (14), the decreasing effect of β with increase in load will compensate for the shift obtained in the da/dt versus J* relation and a single master curve can be obtained for all load conditions.

Some tests have also been conducted under deformation controlled condition (5,15) in which the load P and the deformation rate are controlled in contrast to only one controlled variable, namely, load in creep tests where the deformation rate is allowed to attain its own natural level. The load and the crack length both increase till P reaches P_{max} : then P starts decreasing as $\underline{a}$ increases further. In a typical case (15) the increase in P upto P_{max} is nearly for 3/4 th of the total test period for 304 type stainless steel at $595^{\circ}C$. These deformation controlled experiments are similar to normal tensile testing with a very slow strain rate. Results of deformation controlled tests can represent that of load controlled creep tests only when the deformation rate is purely dependent on the current values of crack growth rate and load and should be independent of the load history. In creep brittle materials the deformation rate will be dependent only on the crack growth rate and not on load (possibility (b)). In such a case, the deformation controlled test can be similar to the load controlled test in which $\Delta \propto (\dot{a})$. A typical example is the Discaloy tested by Landes and Begley (5). In creep ductile materials deformation controlled tests cannot represent load controlled creep tests and so no attempt has been made here to analyse the data obtained under deformation controlled conditions.

CONCLUSIONS

A critical assessment of the experimental data on $\dot{a}$ vs J* relation reveals systematic scatter and load dependence. An analysis based on the energy considerations and the isochro - nous load - deflection curves leads to a parameter R which appears to describe well the creep crack growth data obtained on a ductile material. The parameter clearly distinguishes the behaviour of creep brittle materials (with $\alpha = 0$) and creep ductile materials (with α having a positive value). The parameter is also able to explain the reasons for the systematic scatter and the load dependence of the $\dot{a}$ vs J* relation.

ACKNOWLEDGEMENT

The author expresses his sincere thanks to Prof.Dr. A.J. McEvily, Metallurgy Department, UCONN, Connecticut, USA, for his useful discussions and constant encouragement. Thanks are also due to Prof. P.V. Indiresan, Director, IIT, Madras, for his kind permission to present this paper.

REFERENCES

1. Neate, G.J.   Engg Fracture Mech., 9 (1977) p.297

2. Nikbin, K.M., Webster, G.A. and Turner, C.E.
                        A.S.T.M. STP 601 (1976) p.47

3. Nikbin, K.M., Webster, G.A. and Turner, C.E.
                        Proc ICF-4, (1977) p.627

4. Sadananda, K. and Shahinian, P. Proc 2nd Inter Symposium
                        Elastic-plastic Fracture Mechanics.
                        ASTM, Phil. USA. (1981)

5. Landes,J.D. and Begley,J.A. A.S.T.M. STP 590 (1976)p.128.

6. Koterazawa, R. and Mori,T. Trans A.S.M.E., JEMT
                        99 (1977) p. 298

7. Hutchinson, J.W., Needleman, A. and Shih, C.F.
                        MECH -6, Division of Applied Science,
                        Harward University. (May 1978).

8. Taira, S., Ohtani, R. and Kitamura, T. Trans A.S.M.E.,
                        JEMT, 101 (1979) p.154

9. Sadananda,K. and Shahinian, P. Met Trans., 9(1978)p.79

10. Haigh, J.R.   Mat Sci Engg., 20 (1975) p.213

11. Radhakrishnan, V.M. and McEvily, A.J.
                        Scripta Met., 15 (1981) p.51

12. Sadananda, K. and Shahinian, P.
                        A.S.T.M., STP 791 (1983) p.II-182

13. Radhakrishnan, V.M. and McEvily, A.J.
                        2nd Inter Symposium on Elastic-
                        plastic Fracture Mechanics.
                        ASTM, Phil. USA. (1981).

14. Radhakrishnan, V.M. and McEvily, A.J.
                        Scripta Met., in press

15. Saxena, A.   A.S.T.M., STP 700 (1980) p. 131

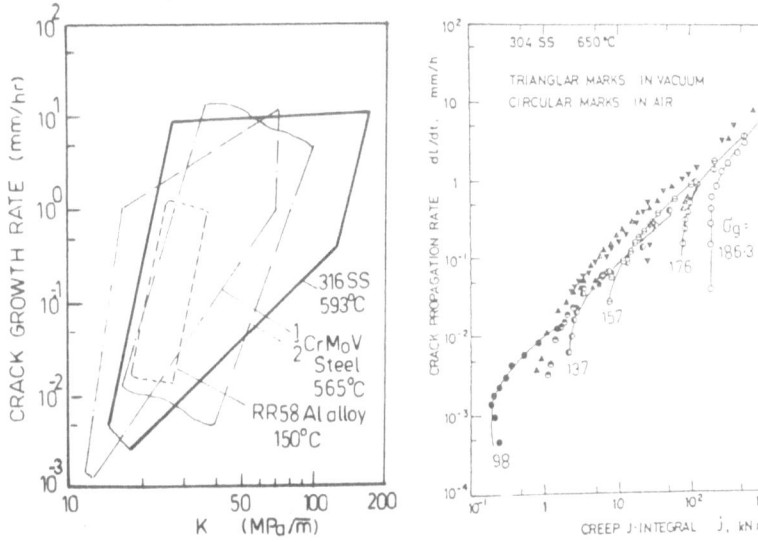

Fig.1. da/dt vs K as represented by scatter band.

Fig.2. da/dt vs J* for
304 type stainless steel.

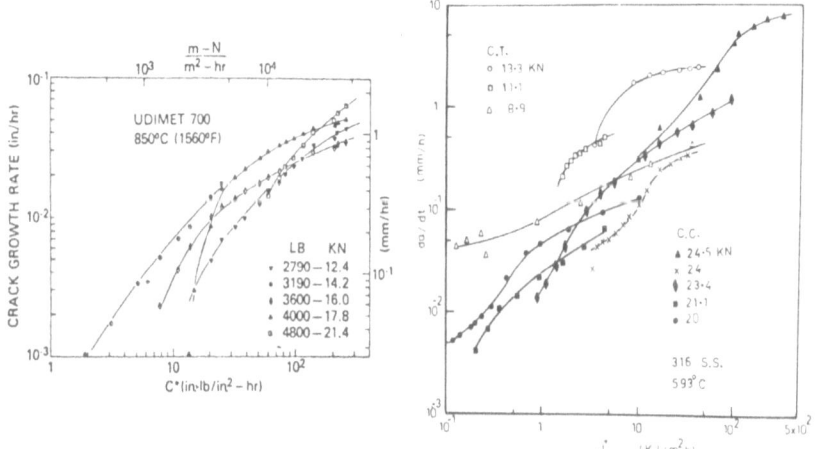

Fig.3. da/dt vs J* for
Udimet 700.

Fig.4. da/dt vs J* for
316 type stainless steel.

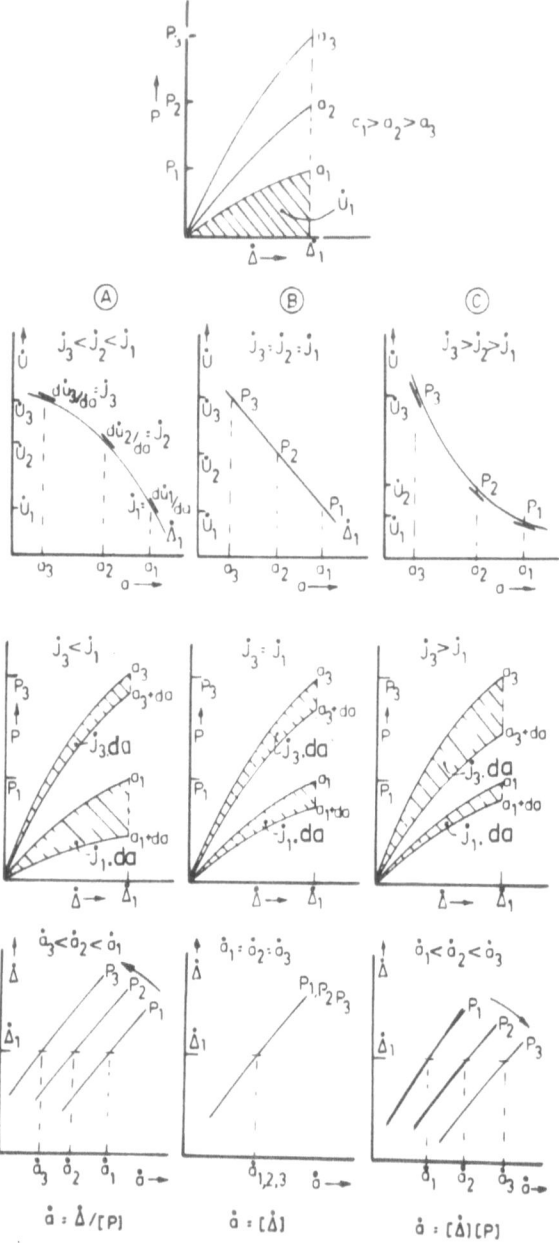

Fig.5.   Data reduction procedure.

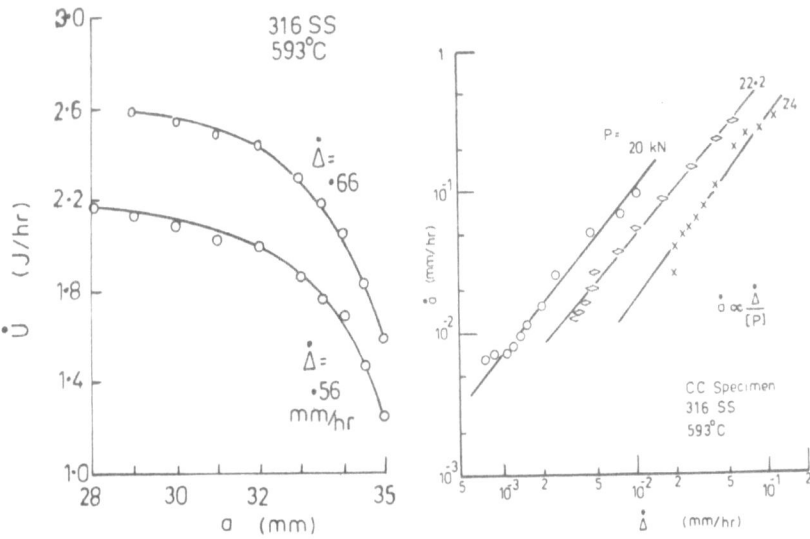

Fig.6. Variation of $\dot{U}$ with crack length.

Fig.7. Relation between $d\Delta/dt$ and da/dt.

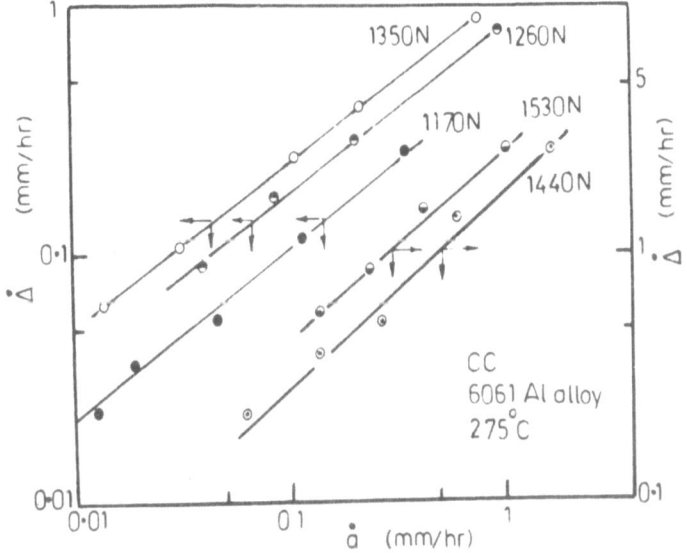

Fig.8. Relation between $d\Delta/dt$ and da/dt for 6061 Al alloy.

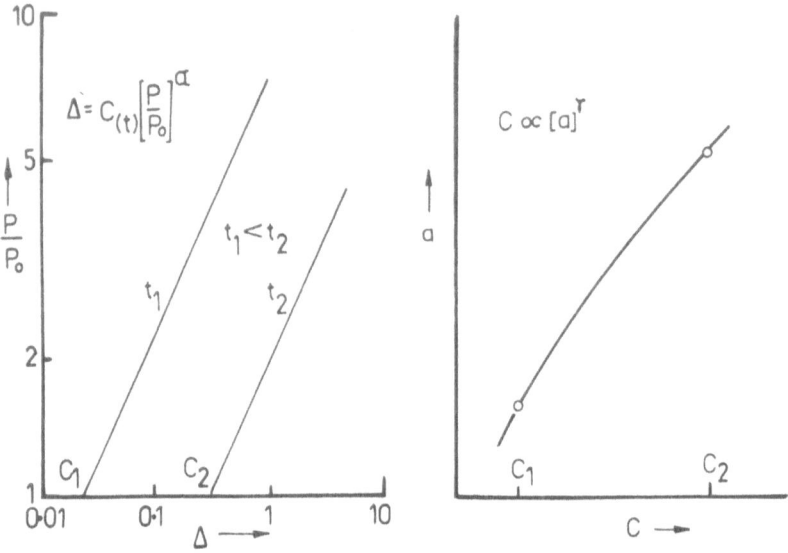

Fig.9. Isochronous load-
deflection curve.

Fig.10. Relation between C
and crack length.

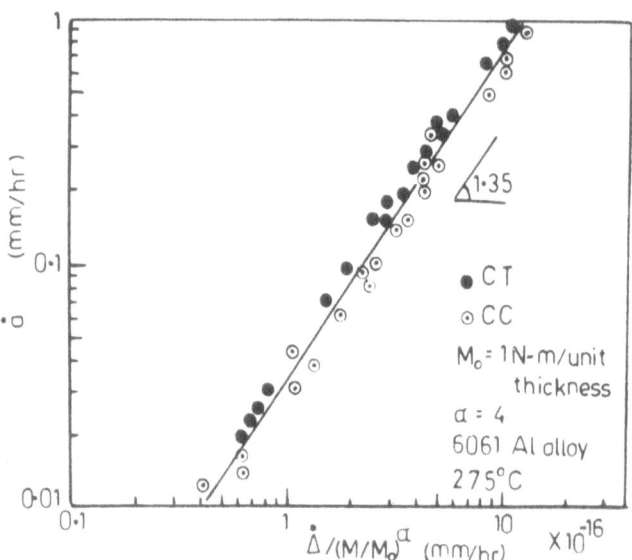

Fig.11. Relation between creep crack growth rate
and the Parameter R.

TIME DEPENDENT PARAMETERS IN DYNAMIC FRACTURE

H. Shimizu*, W.M. Gibbon** and J. Sullivan***

Ontario Research Foundation
  * Research Scientist
 ** Assistant Director, Materials Division
*** Assistant Research Scientist

ABSTRACT

It is usual in instrumented fracture tests, such as Charpy
V-Notch (CVN) and Drop Weight Tear Tests (DWTT), to consider the
total energy expended in the test (E_T), which consists of the energy
expended in achieving peak load (E_i), and the difference between
these two energy values (E_p). The specific details of the sample
geometry are important for each of these energy values. It is
known that none of the CVN energy values is particularly sensitive
to the ductile to brittle transition behaviour of the materials. In
this paper, it is shown that the ratio E_i/E_p is not strongly dependent on the sample geometry but it is sensitive to the type of
fracture. Results for the ratio E_i/E_p from both CVN and DWTT are
found to be in good numerical agreement with one another, and both
results clearly demonstrate the ductile to brittle transition behaviour. Theoretical analysis indicates that this ratio is primarily dependent on crack propagation time.

1. INTRODUCTION

Time dependence in fracture problems usually refers to crack
nucleation and growth processes occurring under low stress conditions
and long incubation periods. Impact loading and dynamic fracture
phenomena are sometimes regarded as a separate category of fracture
processes, because of the large difference in the relative time
frame associated with the static and dynamic fracture processes.
However, close examination of the dynamic fracture behaviour of
metallic materials reveals that the dynamic fracture process also
consists of a sequence of events characterized by time dependency.
Fracture is preceeded by elastic and plastic deformation, including

crack nucleation and growth which ultimately leads to crack
instability and fracture. The major difference in the time
dependent behaviour between static and dynamic fracture occurs in
the crack propagation stage of fracture. Under dynamic loading,
the time required to propagate a crack through a material is orders
of magnitude shorter than in static loading. Quantitative understanding of the dynamic fracture behaviour is made difficult because of the special instrumentation requirements for monitoring
the fracture process (1).

This investigation was undertaken to examine the dynamic
fracture behaviour of linepipe steel in detail using instrumented
drop weight testing of Charpy V-Notch (CVN) and Drop Weight Tear
Test (DWTT) specimens. The primary advantage of instrumented
dynamic testing is its ability to provide data on the instantaneous
load and energy variations during the fracture process.

2. EXPERIMENTAL PROCEDURE

A variety of steel pipe samples exhibiting a similar range of
mechanical properties (Table 1) was selected and subjected to
testing to evaluate their ductile to brittle transition behaviour.
The speciamens for DWT testing were prepared from the two opposite
sides of each pipe as shown in Figure 1, while CVN specimens were
taken from a half section of fractured DWTT specimen. In all cases,
the crack direction was made parallel to the long axis of the pipe.
The full pipe wall thickness (8mm) was retained in all DWTT and CVN
specimens, therefore the CVN specimens conformed to 2/3 size or
subsize Charpy specimens. However, the ligament length was 8mm
with a 2mm depth of V-notch as in the regular CVN specimen. For
temperatures other than the ambient temperature, the specimens were
immersed in a liquid bath of dry ice and alcohol before testing.
The compositions of the four pipes subjected to testing are shown in
Table 2. Impact testing was carried out using a Dynatup instrumented drop weight machine (8000 J capacity).

TABLE 1. Tensile Properties of Pipes

Pipe No.	Yield Stress (0.2%) MPa	UTS MPa	Elongation %	% Area Reduction
A	381	561	33	53.4
B	419	555	33	51.1
C	397	547	32.5	46.4
D	440	601	32.5	39.8

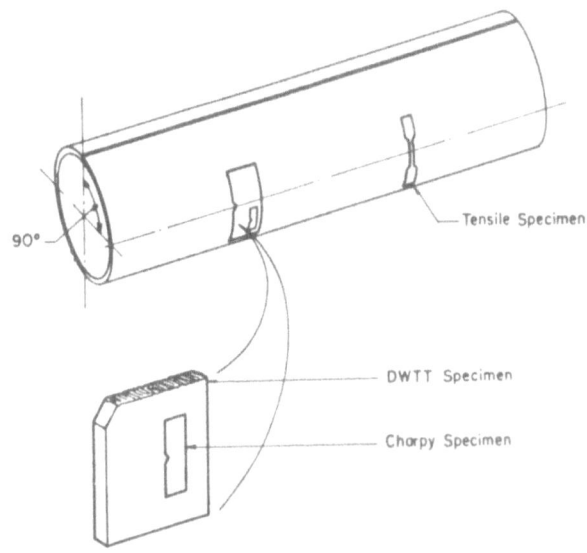

FIGURE 1. Specimen Positions and Orientations

3. RESULTS

The results of impact testing on pipe A are tabulated in Tables 3 and 4 for DWTT and specimens, respectively. From the output traces, such as those shown in Figures 2 and 3, the so-called "initiation energy" E_i at the peak load was calculated. The observed total energy E_T is then related to the propagation energy E by a relationship $E_T = E_i + E_p$.

As expected, the CVN total impact energy of all pipes decreased continuously and smoothly over the entire test temperature range as exemplified by a plot of E_T from CVN testing of Pipe C, shown in Figure 4. It was also expected that the corresponding DWTT results would indicate a much more clear transition behaviour from a ductile

TABLE 2. Chemical Composition

Pipe No.	Si	S	P	Mn	C
A	0.02	0.024	0.011	1.09	0.281
B	0.05	0.012	0.013	1.09	0.226
C	0.01	0.029	0.016	1.12	0.288
D	0.05	0.018	0.010	1.01	0.238

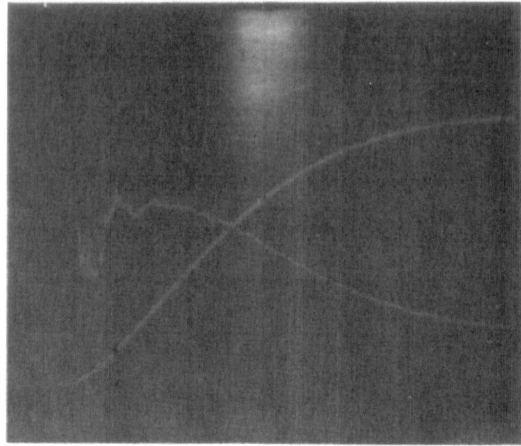

Pipe C

Test Temp.   46°C
Energy Scale 272J/div.
Load Scale   4445N/div.
Time Scale   0.5 ms/div.
Tup Vel.   5.5 m/s

FIGURE 2.   Output traces from DWT testing.

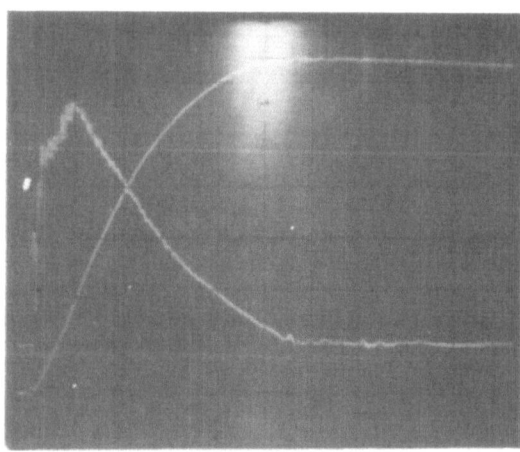

Pipe C

Test Temp.   46°C
Energy Scale 68J/div.
Load Scale   2223N/div.
Time Scale   0.5 ms/div.
Tup Vel.   3.5 m/s

FIGURE 3.   Output traces from CVN testing.

TABLE 3. Summary of Instrumented DWTT Results for Pipe No. A.

Temperature °C	E_i J	E_p J	E_T J	E_i/E_p
46	166	1064	1230	0.16
	175	1068	1243	0.16
35	230	842	1072	0.27
	192	1053	1245	0.18
28	188	922	1110	0.20
	189	574	763	0.33
22	256	277	533	0.92
	203	635	838	0.32
15	107	106	213	1.01
	135	132	267	1.02
-5	167	20	187	8.20
	107	53	160	2.03

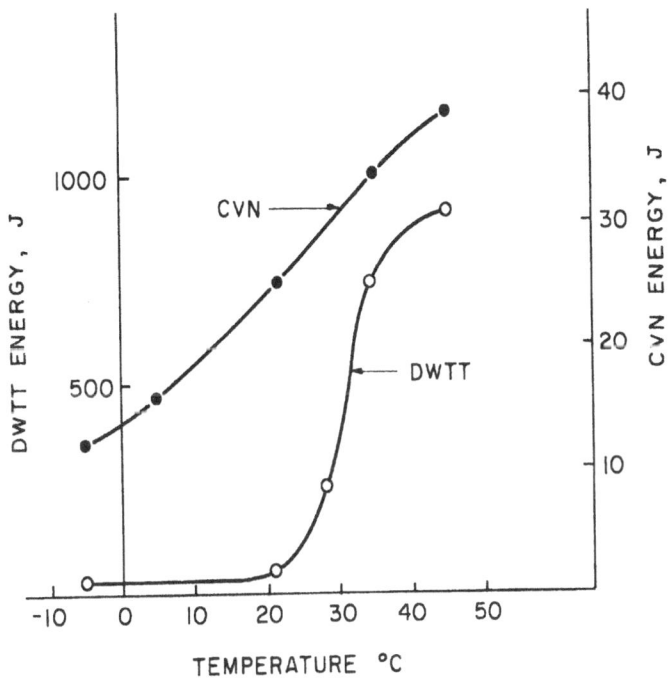

FIGURE 4. DWTT and CVN Energy vs Temperature

228

TABLE 4. Summary of Instrumented CVN Results for
Pipe No. A

Temperature	E_i	E_p	E_T	E_i/E_p
°C	J	J	J	
46	17.5	27.7	45.3	0.63
	14.8	30.5	45.3	0.49
	16.9	25.4	42.3	0.66
34	13.5	20.0	33.5	0.67
	14.1	18.0	32.1	0.79
	11.4	19.3	30.7	0.59
22	10.1	16.7	26.8	0.60
	13.5	13.3	26.8	1.01
	16.2	15.2	31.4	1.06
5	10.7	6.9	17.6	1.55
	10.7	6.7	17.4	1.61
	12.5	7.6	19.6	1.64
-5	10.7	4.2	14.9	2.55
	10.7	4.2	14.9	2.55
	11.7	4.0	15.7	2.90

to a brittle mode of fracture. The E_T values from DWTT for Pipe C are also shown in Figure 4. Other traditional measure of ductility such as the shear area and lateral expansion values varied in much the same way with temperature as the total energy values from the two test techniques.

4. DISCUSSION

The charpy test is the oldest test technique which is in general use for the evaluation of fracture resistance of materials. Numerous attempts have been made to correlate the service performance of structures or large components from some criteria based on the Charpy test. Some notable examples are cited in the references (2, 3, 4 and 5). Experience has shown that successful correlations are more of an exception than a general rule, and they are applicable within narrow limitations of materials and environmental conditions. The difficulty of correlation is related fundamentally to the geometry of the test piece, and not to any deficiencies in the choice of critera.

The difficulty stems from the fact that the amount of ductility that is exhibited by a specimen is dependent on its geometry. There are three main problems related to the Charpy specimen geometry: (1) the specimen width is restricted to 10mm, to which thicker materials must be machined down, (2) the limited ligament length of the specimen, and (3) the radius of the notch is relatively dull compared with natural cracks. When such a specimen is subjected to three-point bending, general yielding of the entire ligament section is possible before crack initiation can occur and thus the total energy measured in fracturing the specimen contains a large proportion of initial deformation energy.

The DWTT specimens were designed primarily to obtain information related to fracture surface appearance. The test was orininally adopted in pipeline industries for testing full thickness plate materials, and to evaluate their ductile to brittle transition based on the amount of shear fracture in fractured specimens. It has been regarded that energy measurements from the DWTT are not meaningful because the small notch causes the ligament to deform plastically prior to fracture, and leads to the same problem as described above.

Such traditional objections to the limitation imposed by specimen geometry considerations are valid when the fracture energy measurements represents only the total energy to fracture a speciment. Instrumented impact testing provides additional critical information on the initial deformation characteristics of the test specimen. From the viewpoint of analyzing the elastic vs plastic deformation components, it is not essential that the E_i value corresponds to the crack initiation energy. It is important that a consistent reference point is chosen to define the initial deformation region. For this purpose, the energy at the peak load is a convenient and recognized reference point.

It is therefore reasonable to examine the total energy value in terms of the elastic and plastic components in the fracture process, and to evaluate their relative contribution to the overall fracture energy for the two types of specimens.

Close examination of these energy components in both CVN and DWTT fracture revealed that the ratio, E_i/E_p, over the test temperature range may be a much more sensitive indicator of ductile to brittle transition behaviour than the total energy or component energy values. The results shown in Figure 4 were reanalyzed in terms of the E_i/E_p ratios as in Figure 5, which shows the results of DWTT samples taken from two opposite regions of a pipe and the instrumented CVN (ICVN) results obtained from the same samples.

It is apparent that the two test methods yield essentially equivalent results when treated in this manner. The application of this technique to other pipes confirmed that the CVN and DWT testing can be correlated, and in some cases, nearly exactly.

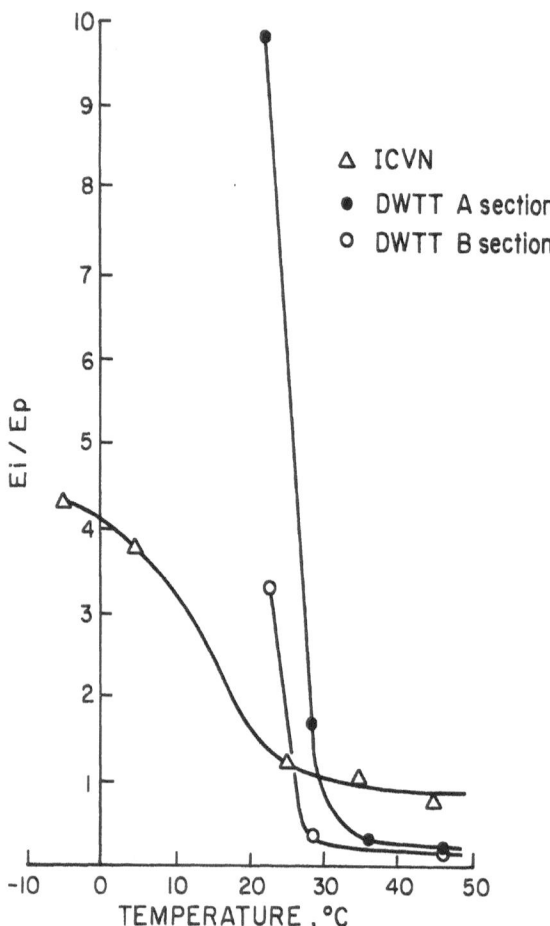

FIGURE 5. Ratios of E_i/E_p vs temperature

These results suggest that in spite of the large difference in the specimen sizes, the ductile to brittle transition of this material is virtually independent of geometry when comparisons are made in terms of the energy ratios involved in the respective phases of the fracture processes: initiation and propagation phases. Figure 5 shows that the onset of brittle fracture is reflected in the increasing values of the ratio E_i/E_p (or E_i/E_T) as the material enters the "initiation" dominated temperature regime as opposed to the "propagation" dominated temperature regime. These results are consistent with the observation that in practise, brittle fracture is characterized by the absence of plastic deformation.

The complications of the energy ratio analysis are being examined theoretically and preliminary results are presented below. It is possible to express the two stages in terms of (1) the shear displacement equations and (2) the known fracture toughness equations; the latter being used to approximate the dynamic propagation stage as a series of quasi-static events. Such an analysis leads to a conclusion that the ratio (E_i/E_p) can be represented by:

$$\frac{E_i}{E_p} = \frac{K_{IC}}{24\ YG}\ \frac{1}{a_0}\left[\frac{Us}{As}\sqrt{\frac{\pi}{8E\gamma_s}} + \sqrt{a_0}\right]\left[\frac{W}{V_T\ t_f}\right]$$

where

K_{IC}	=	Critical plain strain fracture toughness
Y	=	a geometrical constant
G	=	the shear modulus
a_0	=	the notch depth
Us	=	the energy to form a fractured surface
As	=	the corresponding fracture area
E	=	the elastic constant
γ_s	=	the surface energy
W	=	the depth of the specimen
V_T	=	the tup velocity
t_f	=	the time required for crack propagation

Therefore, the energy ratio is basically a product of four terms:

$$\frac{E_i}{E_p} = C\ K_{IC}\left(\frac{Us}{As}\right)\left(\frac{W}{V_T\ t_f}\right)$$

The first term, C, is a constant which depends only on the material and the second term, K_{IC}, is dependent on fracture mode. The last term is almost self compensating for the different ligament length, and the third term is nearly invariant so long as the fracture mode remains ductile and unchanged. However, as the material becomes brittle the fracture area As and fracture time t_f show a significant decrease, leading to an increase in the value of (E_i/E_p). It seems therefore that this ratio is a sensitive indicator of ductile to brittle transition which is essentially independent of the geometry of the specimen.

5. CONCLUSION

It has been demonstrated that instrumented impact testing can provide some basic data relevant to the ductile-to-brittle transition behaviour of a material. It is suggested that the fundamental difficulty exists in interpreting the total fracture energy of any fracture testing methodology because of the difficulty of separating the elasto-plastic bending phase from the crack propagation phase of the fracture process. The energy requirements for the two phases are dominated by the compliance factor in the former phase and by the microstructural factors of the material in the latter phase, resulting in an irresolvable dependence of fracture energy on specimen geometry.

It is suggested that this difficulty may be partially resolved by examining the individual energy terms and by comparing their ratio. The results of this investigation indicate that the fracture area and fracture times are two of the indicative parameters which signal the onset of brittle fracture behaviour.

6. REFERENCES

1. Shimizu, H. and W.M. Gibbon. Evaluating the Dynamic Toughness Properties of Pipeline Steels. Canadian Metallurgical Quarterly 21 (1982) 103-109.

2. Impact Testing of Metals. Special Technical Publication 466 (Philadelphia, American Society for Testing and Materials, 1970).

3. Irwin, G.R. and R. Roberts. Fracture Toughness of Bridge Steels. Phase I. (Bethleham, Pa., Report, Lehigh University, 1972).

4. Eiber, R.J. and W.A. Maxey. Materials Engineering in the Arctic. Proceedings of an International Confernece, St. Jovite, Quebec, Canada. (Metals Park, Ohio, American Society for Metals, 1977) 306-319.

5. Deeks, R.G. Materials Engineering in the Arctic. Proceedings of an International Conference, St. Jovite, Quebec, Canada. (Metals Park, Ohio, American Society for Metals, 1977) 110-116.

A PROPOSED RATIONALE FOR ACCOUNTING FOR BOUNDARY LAYER EFFECTS IN DESIGNING AGAINST FRACTURE IN THREE DIMENSIONAL PROBLEMS

C. W. Smith, O. Olaosebikan and J. S. Epstein[a]

Department of Engineering Science & Mechanics
Virginia Polytechnic Institute & State University
Blacksburg, Virginia 24061
a- Currently at Oxford University

ABSTRACT

After briefly reviewing the features of an optical method for measuring boundary layer effects in cracked body problems, results of its application to two problem classes are presented. An interpretation of these results leads to a suggestion for accounting for boundary layer effects in design rationale.

1. INTRODUCTION

Perhaps the major generic problem in the realm of stable crack growth can be described as the initiation of a crack from a defect in the neighborhood of a stress raiser which grows under cyclic or repeated mechanical and/or thermal load until failure results. The stable growth regime often involves curved crack fronts, non-planar crack surfaces and complex boundary conditions, all of which lead to a non-uniform stress intensity factor (SIF) distribution along the crack front. Such complications force analysts to use numerical methods of analysis, often without adequate experimental computer code validation.

Beginning over a decade ago, the first author and his associates began a study directed towards the development of experimental modelling methods which could be employed to provide code verification for such models. Originally proposed for Mode I only [1] the methods have been extended to include all three modes of near tip analysis [2] photoelastically and currently also include auxiliary moire interferometric determination of displacements [3].

In 1970, G. C. Sih [4] and later E. S. Folias [5] focused attention on the crack border-free surface intersection problem. Then, following two recent landmark papers by Prof. J. P. Benthem [6],[7], it has been verified both analytically [8],[9],[10] and experimentally [11] that the inverse square root singularity commonly assumed in classical fracture mechanics is lost when a crack intersects a free boundary at right angles. In thick bodies of incompressible materials, this effect may alter the stress intensification near the boundary by almost a factor of two but a smaller effect is even felt remote from the intersected boundary.

Using the ASTM E-399 Compact Bending Specimen Geometry as a model, the authors have been able to measure the order of the singularity and its variation through the beam thickness using high density moire interferometry. Results for incompressible materials, and also for a material with Poisson's Ratio ≈ 0.4, agree with Benthem's results to well within experimental error. However, these results raise the issue of how the designer is to apply this knowledge to classical fracture mechanics in a rational design process.

After briefly reviewing the experimental methods and algorithms for converting data into stress intensity parameters, results obtained from two specific test configurations, namely the ASTM E-399 Compact Bending Specimen and part through cracks under Mode I loading are presented. The compact bending results are used to provide benchmark data and the part through cracks are used to indicate the effect of changing the boundary intersection angle as observed when fatigue or creep crack growth occurs. Finally, a "corresponding" stress intensity factor (K_{cor}) is proposed for use by designers as a means of accounting for the complexities of the boundary layer effect in cracked bodies of substantial thickness.

2. EXPERIMENTAL METHODS

As noted above, two experimental methods, frozen stress photoelasticity, and high density moire interferometry have been utilized in studying the above noted boundary layer effect as well as the variation in the SIF distribution in three dimensional cracked body problems. These methods have been described in detail elsewhere [2],[12] so will only be briefly reviewed here.

2.1 Frozen Stress Photoelasticity

This method capitalizes on the fact that the transparent model material exhibits diphase mechanical and optical behavior. At room temperature, it responds to load in a Kelvin-like manner (Fig. 1). However, when the model's temperature is raised to its

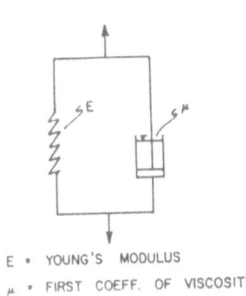

E • YOUNG'S MODULUS

μ • FIRST COEFF. OF VISCOSITY

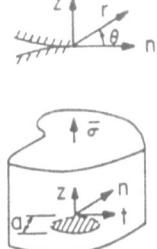

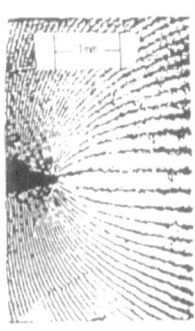

Fig. 1 Model of Kelvin Material

Fig. 2 General Problem Configuration and Notation

Fig. 3 Moire Displacement Fringes for U_z (Mode I)

"critical" value, the viscous coefficient vanishes and the mechanical response becomes linearly elastic with a corresponding reduction in material modulus of some two and a half orders of magnitude and an increase in optical sensitivity to stress fringes of about twenty five times the room temperature sensitivity. To take advantage of this behavior, we heat the cracked model to critical temperature, load and cool under load. Upon load removal, and even after slicing through the body after load removal, the deformations and stress fringes produced above critical temperature are retained and can be analyzed photoelastically in the usual way. The photoelastic fringes are proportional to the maximum shearing stress in the plane of the slice, and the algorithms for converting the data into stress intensity information are referred to a local cartesian coordinate system tnz (Fig. 2) with slices always taken parallel to the nz plane for Mode I loading [1],[2].

2.2 High Density Moire Interferometry

By assuming that the annealing of stress frozen slices reverses the effect of stress freezing (which has been verified experimentally) one can deposit a grating of 1200 lines/mm on a slice surface and then anneal the slice, deforming the grating. By viewing this grating through a "virtual" master grating developed by Post [13], moire fringes which are proportional to the in-plane displacement components normal to the master grating can be obtained. The virtual grating is formed by splitting and recombining rays from a laser light source so as to form an optical grating of lines of constructive and destructive interference. It is important that rigid body motions be eliminated from the data used to compute SIF values. By following the photoelastic analysis of a slice with a moiré analysis, two separate estimates of the SIF for that slice may be obtained.

3. ALGORITHMS FOR CONVERTING DATA INTO SIF VALUES

With the measuring techniques described above, we measure maximum shearing stress magnitude and displacement components in the slice plane. Generally, these data are collected close to the crack tip, but must exclude a very near tip non-linear zone. For problems in which boundary layer effects are not present (such as embedded flaws) or are neglected, algorithms are constructed from classical linear elastic fracture mechanics (LEFM) for converting optical data into SIF values. These algorithms have been dis-, cussed elsewhere [2,12] and will not be repeated here. In order to study the boundary layer effect, however, algorithms must be constructed which are outside the realm of LEFM.

3.1 Moiré Displacement Algorithm

Benthem's solution (loc. cit.) was a variable separable result which, for the displacement components, took the form of a series of eigenfunctions for a quarter infinite crack in a half space. Focusing on the lowest order eigenvalue term as is customary in such problems and noting that, for Mode I, moiré fringes (Fig. 3) are readily measurable along $\theta = \pm \pi/2$ we have near the crack tip:

$$u_z = Cr^{\lambda_u} \tag{1}$$

u_z = component of displacement normal to crack plane
C = coefficient of leading term in displacement series, constant for $\theta = \pm \pi/2$
λ_u = first or lowest order eigenvalue in solution for displacement

$$\therefore \log u_z = \log C + \lambda_u \log r \tag{2}$$

and λ_u may be determined as the slope of a plot of $\log u_z$ vs $\log r$. Fig. 4 shows a typical result from a slice removed from a compact bending specimen. The $\log r$ location for the slope measurement is in the linear range with the smallest slope and is common to all slices for a given crack border. In general the zone lies between $r = 0.1$mm and 1.0mm from the crack tip.

3.2 Photoelastic Algorithm

Since data from a stress frozen slice is averaged optically through the slice thickness, we can describe this data locally with a two dimensional function analagous to a Westergaard stress function for a small enough data zone taken so as to focus upon

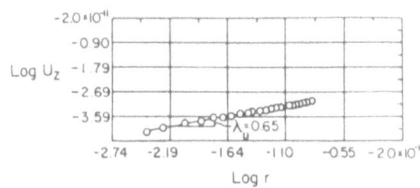

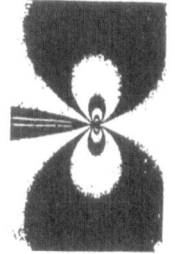

Fig. 4 Determination of λ_u from Moire Data

Fig. 5 Mode I Isochromatics Near Crack Tip

the in-plane singular effect. Using this approach, and again focusing upon the lowest order eigenvalue, we have, along $\theta = \pm \pi/2$ where the stress fringes spread the most for Mode I (Fig. 5)

$$\lim_{|\zeta| \to 0} [Z(\zeta)] = \frac{K_\lambda}{\sqrt{2\pi}\ r^{\lambda_\sigma}} \qquad (3)$$

where

Z is the stress function
ζ is defined in Fig. 6
K_λ is a "stress eigenfactor"
λ_σ is the lowest order eigenvalue in the stress equations.

By computing σ_{ij} (i,j = n,z) from Eq. (3) and τ_{nz}^{max} from these stresses, one arrives at the approximate expression [14]:

$$\tau_{nz}^{max} = \frac{K_\lambda f(\lambda_\sigma)}{r^{\lambda_\sigma}} + \sigma_{on} \qquad (4)$$

where σ_{on} represents the contribution of the nonsingular stresses in the measurement zone. By defining:

$$(K_\lambda)_{AP} = \tau_{nz}^{max}\ r^{\lambda_\sigma} \qquad (5)$$

then

$$\log(\tau_{nz}^{max}) = \log(K_\lambda)_{AP} - \lambda_\sigma \log r \qquad (6)$$

Thus, by plotting $\log(\tau_{nz}^{max})$ vs log r, λ_σ can be determined. We note that $\lambda_\sigma = \lambda_u - 1$. However, they can be determined independently of this relation.

238

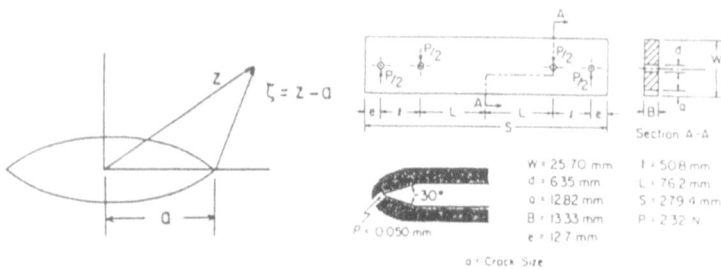

Fig. 6   ζ Coordinate   Fig. 7   Four Point Load Compact
                                        Bending Specimen

4.1 Results from Compact Bending Experiments

The test specimen for these experiments is given in Fig. 7.
Displacement fields obtained for each slice by moire interferometry using the procedures of Sec. 2.2 were converted through the
algorithm described in Sec. 3.1 using Eq. (2) into values of λ_u.
Stress fields obtained photoelastically were converted in a similar way using the procedures of Sec. 2.1 and the algorithm of Sec.
3.1 using Eq. (6). Since Benthem's solution was for a half space,
we compared our results with his only at the free surface. They
were as follows:

Poisson's Ratio	λ_u (Expr.)	λ_u [7]
0.40	0.58	0.59
0.48	0.63	--
0.50	--	0.65

Once our experimental results were benchmarked in this way, we
then plotted the variation in λ_u across the beam half thickness
for both moire and photoelastic data. For the photoelastic data
$\lambda_u = \lambda_\sigma + 1$. The result is shown in Fig. 8. We conclude that the
methods yield reasonable results with good accuracy.

4.2 Results from Part Through Crack Experiments

Benthem's solution and the compact bending results presented
in the previous section are for straight front cracks which intersect free surfaces at right angles. When cracks grow under fatigue or creep, the boundary angle appears to change. In order to
study the boundary layer effects for other boundary intersection
angles, part through cracks were employed. Fig. 9 gives the distribution of λ_σ for such a geometry. However, in these studies,
no moire data were taken on the free surface itself. Instead,

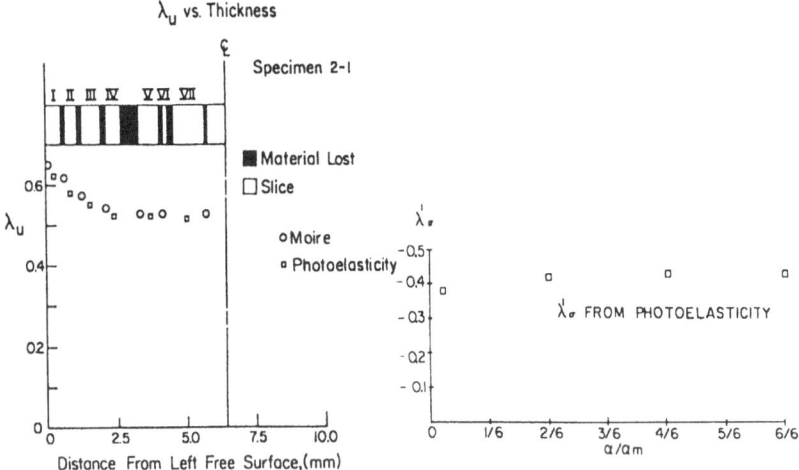

Fig. 8 λ_u Values from Moire and Photoelastic Data

Fig. 9 λ_u Distribution from Part Circular Flaw

slices positioned as shown in Fig. 10a were progressively reduced in thickness by sanding the surface furtherest from the free surface through which the crack entered. As shown in Fig. 10b, there is a severe reduction in the classical SIF as the slice thickness approaches zero. In fact, as shown by the dashed lines, the classical SIF could even vanish at the free surface.

5. DISCUSSION

In the foregoing, experimental methods of analysis were used to measure the variation in the order of the lowest eigenvalue for cracks intersecting free surfaces. The effect is greatest for materials with high Poisson's ratio (rubbers, plastics, adhesives, rocket motor propellants) and, for such materials one raises the question as to how the effects should be accounted for by the fracture analyst in designing against fracture. Since K_λ, or $K_\lambda f(\lambda)$ do not have dimensions of the classical SIF, some additional interpretations are needed here.

When the frozen stress method is employed, the quantity measured is proportional to a stress averaged through the thickness in a local region near the crack tip. In this region we have obtained expressions for an apparent stress intensity value of both LEFM [3] and from Eq. (5) for the boundary layer algorithm. Since each describes the same quantity (in-plane maximum shear stress) we equate them in the measurement zone to obtain:

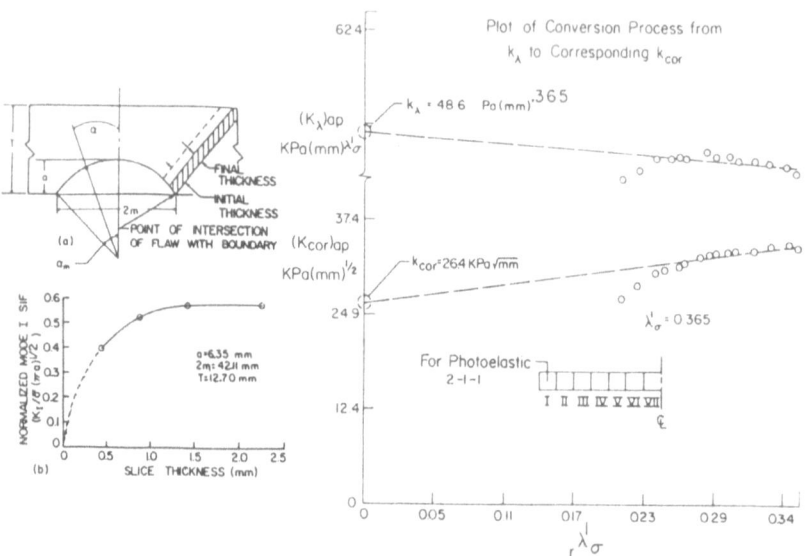

Fig. 10   (a) Thinning of Part
Through Crack Slice
Which Intersects the Free
Surface. (b) Variation
of Classical SIF with
Slice Thickness (see
Fig. (10a)

Fig. 11   Converting from $(K_\lambda)_{AP}$
$(K_{cor})_{AP}$

$$\tau_{nz}^{max} = \frac{K_{AP}}{r^{1/2}} = \frac{(K_\lambda)_{AP}}{r^{\lambda_\sigma}} \qquad AP \to \text{"apparent"} \qquad (7)$$

where K_{AP} "corresponds" to $(K_\lambda)_{AP}$ so we call it $(K_{cor})_{AP}$. Thus

$$(K_{cor})_{AP} = (K_\lambda)_{AP} \ r^{1/2-\lambda_\sigma} \qquad (8)$$

If we then use Eq. (8) on the boundary layer data, we can convert
to a "corresponding" classical LEFM value. Once this is done,
$(K_{cor})_{AP}$ values may be extrapolated to the origin on a plot of K_{AP}
vs r to obtain K_{cor} values as is done to obtain K_I in LEFM [12].
Fig. 11 shows an example of this procedure. In Fig. 12 we compare
the photoelastic data using an LEFM algorithm with the results
obtained using the boundary layer algorithm and Eq. (8) for the
compact bending specimen. We note that use of Eq. (8) essentially
eliminates the influence of the free surface although the result
is slightly higher than the two dimensional result. Thus the new

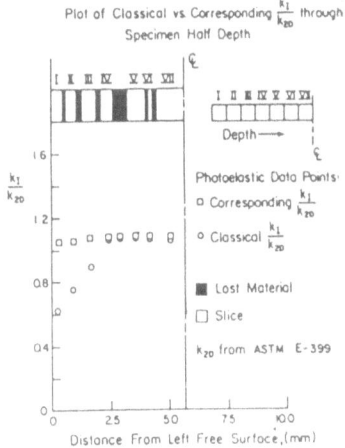

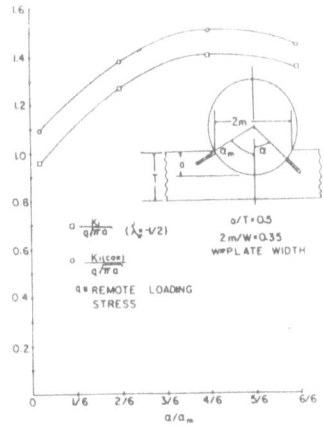

Fig. 12 K_I and K_{cor} Distributions in Compact Bending Specimens

Fig. 13. K_I and K_{cor} Distributions in Part Through Crack in a Wide Plate

algorithm, when modified by Eq. (8), yields a conservative result which could be employed within a two dimensional context by the designer. Fig. 13 shows a similar result for a part through crack. We again see an elevation of K_{cor} above the classical SIF as before. However, a large zone of variation in K_{cor} also appears which was not present for the straight front crack with a right angle free boundary intersection. Through conjecture, we now address the reason for the difference in the K_{cor} distributions in Figs. 12 and 13. If we conjecture that, when the 90° intersection angle is altered during crack growth, the inverse square root singularity is being restored near the boundary, then the difference between the K_I and K_{cor} distributions in Fig. 13 should be small and, perhaps similar in shape. This is the effect we observe. Conversely, if the singularity is completely lost at the boundary in the limit as the slice thickness goes to zero, then this could also account for the larger gradient of both K_I and K_{cor} in Fig. 13. Another issue involves the applicability of the algorithm to the part through crack. Studies directed towards these issues are continuing. One might argue that it would be preferable to equate the stresses on the plane of the crack ahead of the crack tip and that approach is also under study. It appears, however, at this writing that much better data are available along $\theta = \pm \pi/2$.

6. SUMMARY

Experimental methods of analysis and corresponding algorithms for converting data into fracture parameters were reviewed. Results obtained from applying the methods to an analysis of boundary layer effects were presented. Finally, a means for incorporating the new results into LEFM design rationale was suggested. Studies relating to this latter point are continuing.

ACKNOWLEDGMENTS

The authors wish to acknowledge the contributions of their colleagues as noted in the text in providing foundations for this work. The support of the National Science Foundation under Grant No. MEA-811-3565 is also gratefully acknowledged.

REFERENCES

[1]   Smith, C. W., "Use of Three Dimensional Photoelasticity and Progress in Related Areas", Experimental Techniques in Fracture Mechanics 2, SESA Monograph No. 2, A. S. Kobayashi, Ed., Ch. 1, 1975, pp. 3-58.

[2]   Smith, C. W., "Use of Photoelasticity in Fracture Mechanics", Experimental Evaluation of Stress Concentration and Intensity Factors, Mechanics of Fracture, Vol. 7, G. C. Sih, Ed., Martinus-Nijhoff, Ch. 2, 1981, pp. 162-187.

[3]   Smith, C. W., "Use of Optical Methods in Stress Analysis of Three Dimensional Cracked Body Problems", J. of Optical Engineering, Vol. 71, No .4, 1982, pp. 696-703.

[4]   Sih, G. C., "A Review of the Three-Dimensional Stress Problem for a Cracked Plate", Intl. J. of Fracture Mechanics, Vol. 7, No. 1, March 1971, pp. 39-61.

[5]   Folias, E. S., "On the Three Dimensional Theory of Cracked Plates", J. of Applied Mechanics, Vol. 42, Series E, No. 3, Sept. 1975, pp. 663-672.

[6]   Benthem, J. P., "On an Inversion Theorem for Conical Regions in Elasticity Theory", J. of Elasticity, Vol. 9, No. 2, 1979, pp. 159-169.

[7]   Benthem, J. P., "The Quarter Infinite Crack in a Half Space: Alternative and Additional Solutions", Intl. J. of Solids & Structures, Vol. 16, 1980, pp. 119-130.

[8]   Bazant, Z. P. and Estenssoro, L. F., "Stress Singularity and Propagation of Cracks at Their Intersection with Surfaces", Northwestern Univ. Struct. Engr. Report No. 77-12/480, Dec. 1977.

[9]   Burton, W. S. and Sinclair, G. B., "On the 3D Implications of LEFM, An Intergral Equation Approach", 16th National Symposium on Fracture Mechanics, Battelle-Columbus, August 1983.

[10] Solecki, J. S. and Swedlow, J. L., "On the 3D Implications of LEFM - Finite Element Analysis of Straight and Curved Through Cracks in a Plate" (see Ref. 9).

[11] Smith, C. W. and Epstein, J. S., "Measurements of Near Tip Fields Near the Right Angle Intersection of Straight Front Cracks" (In Press), Proc. of Tenth Canadian Fracture Conference, Waterloo, Canada, Aug. 1983.

[12] Smith, C. W., Post, D. and Nicoletto, G., "Experimental Stress Intensity Distributions in Three Dimensional Cracked Body Problems", Experimental Mechanics, Vol. 23, No. 4, Dec. 1983, pp. 378-382.

[13] Nicoletto, G., Post, D. and Smith, C. W., "Moiré Interferometry for High Sensitivity Measurements in Fracture Mechanics", Proc. of 1982 joint SESA-JSME Conference on Experimental Mechanics, 1982, pp. 258-262.

[14] Smith, C. W. and Epstein, J. S., "Measurement of Three Dimensional Effects in Cracked Bodies" (In Press), Proc. of Vth Intl. Congress on Experimental Stress Analysis, June 1984.

THE INFLUENCE OF MICROSTRUCTURE UPON THE CREEP AND FATIGUE CRACK GROWTH BEHAVIOUR IN INCONEL 718

R. Thamburaj**, A.K. Koul*, W. Wallace*, T. Terada*,
and M.C. de Malherbe**

*National Research Council, Ottawa
**Carleton University, Ottawa

ABSTRACT

The influence of microstructural variations upon the creep and fatigue crack growth rates in Inconel 718 at 650°C have been studied through a fracture mechanics approach. Crack growth rates have been measured for four different heat treatment conditions, one of which was designed to create a serrated grain boundary structure.

Each microstructural condition has been tested in air and in vacuum to determine the contribution of an oxygen containing environment to elevated temperature crack growth behaviour. A constant-K fracture mechanics specimen has been designed and used for the measurement of crack growth rates. The observed differences in the rates of crack extension have been briefly discussed in relation to the possible differences in the micro-mechanisms of deformation and fracture under each condition.

INTRODUCTION

At high temperatures, fatigue crack growth rates (FCGR) become strongly time and temperature dependent, not only because of recovery and creep effects, but also because of very active environmental effects. Understanding the complex nature of such creep-fatigue-environment interactions is one of the major challenges in high temperature metallurgy today.

The objective of this study is to describe a fracture mechanics specimen which can be used to study time-dependent crack growth behaviour under inert and aggressive environments and to detail some results concerning the influence of microstructure on creep and fatigue crack growth rates in Inconel 718 at 650°C.

PHYSICAL METALLURGY OF INCONEL 718

In the conventionally heat treated condition, Inconel 718 contains the gamma phase (the matrix), a coherent body-centered-tetragonal Ni_3Cb precipitate referred to as gamma double prime (γ'') and also a coherent face-centered-cubic phase known as γ' ($Ni_3(Al,Ti)$). Small amounts of CbC and TiN are present also.

The γ' precipitate usually occurs during aging as uniform, fine particles in the γ matrix. These particles are coherent with the matrix and their solvus temperature is between 843° and 871°C [1,2]. The γ' precipitate contributes only about 10-20% to the hardening of the alloy [2] and transforms to γ'' upon exposure to temperatures around 788° to 815°C for 1h [1,3,4].

Between 650-760°C, the γ'' precipitate normally forms as discs with an average diameter of 600Å and a thickness of 50-90Å [2]. Above 760°C it exhibits accelerated coarsening. γ'' transforms to orthorhombic Ni_3Cb (δ) after exposure at temperatures above 815°C [3,5].

Orthorhombic Ni_3Cb forms as large needles in the grains and at the grain boundaries. Although quite stable, it is not coherent with the matrix. Therefore, it contributes little to the strength and may harm ductility. However, small amounts of Ni_3Cb at the grain boundaries can give rise to a serrated grain boundary morphology and this is believed to improve notch rupture ductility at 650°C [6]. At temperatures above 982°C, the δ phase generally dissolves [7]. Heat to heat variations can cause significant differences in the time taken for δ phase formation [1,3].

EXPERIMENTAL OUTLINE

Material

The material was commercially made 12.7 mm plate stock of Inconel 718. Four heat treatments, described in Table 1, were applied to this material.

The "Conventional" heat treatment results in a fine-grained microstructure (Fig. 1a). The matrix contains a distribution of particles ~.2μm in size as well as a very fine, barely resolvable precipitate (Fig. 2a). The "Modified" heat treatment produces a significantly coarser grain size than the conventional heat treatment (Fig. 1b). The precipitate morphology is shown in Fig. 2b. The coarse particles form during the slow cool to the aging temperature and the finer ones during the aging treatments at 718 and 621°C.

In the treatment suggested by Merrick [6] δ-phase precipitation during the intermediate age is said to lead to an irregular grain boundary morphology. However, in the heat of Inconel 718 used in the present study, this treatment produced very planar grain boundaries (Fig. 1c). The precipitates consist of disc shaped γ" particles 0.2 μm in size as well as spherical particles which are probably γ'.

A modification of Merrick's treatment which employed a furnace cool of ~7°C/min to the intermediate aging temperature produced a significant amount of serration at the grain boundaries (Fig. 1d). These serrations are angular in nature and hence are quite different from the well rounded serrations formed in other Ni-base superalloys [8]. As a result of the furnace cool from the solution annealing temperature and the intermediate age at 843°C, a significant amount of γ" coarsening was observed in this micro-structure (Fig. 2d), several particles being 1 μm in diameter.

Apparatus

The test facility consisted of an Inconel 600 environmental chamber installed in an MTS servohydraulic test system. As illustrated in Fig. 3, the fracture mechanics specimen was held through a load train assembly passing through the retort. Sealing was provided at the top and bottom by welded metal bellows and the retort could be evacuated to a pressure of 2×10^{-6} Torr at temperatures up to 650°C.

Specimen heating was achieved through a three-zone resistance heated split furnace which could be wrapped around the retort. With this arrangement, temperature variations over the length of the specimen were within ±2°C at a temperature of 650°C.

Crack growth rate measurements were made with the aid of a clip-gage type extensometer shown in Fig. 3. This device could be clipped on grooves machined in the specimen and was calibrated to measure crack length as a function of the displacement of the specimen at these grooves. This instrument was found suitable for measuring crack growth rates under static loading conditions and for fatigue at low frequencies (0.1 Hz to 1.0 Hz). Perhaps due to the several mechanical links present, this device was found to have sluggish response to specimen displacement at frequencies above 2.0 Hz.

Specimen

It was desired to use a fracture mechanics specimen that would provide an approximately constant value of the stress intensity factor (K) over a significant range of crack lengths. A constant-K approach has several advantages as listed below.

1. Crack propagation rates are evaluated more easily than in rising-K (or ΔK) tests. For example, creep crack growth rates under constant-K conditions may be determined simply by dividing the crack extension by time under static load.

2. Crack propagation rates can be determined in very aggressive environments which do not permit the use of extensometers or other instrumentation within the retort.

3. Hold times can be applied at constant-K.

4. Environmental effects on crack growth rates are known to depend upon the value of the stress intensity factor [9]. By keeping the value of K constant, the study of the influence of dynamic changes in the environment is facilitated.

Constant-K conditions can be conveniently obtained with a tapered double cantilever beam (DCB) specimen. Though this type of specimen has been used in a number of earlier investigations [10-13], the specimens available so far are quite large compared to the compact tension specimens being used for high temperature crack growth monitoring [14]. They may present difficulties if used for a creep-fatigue environment interaction study of the type proposed here. For example, it may take several weeks to complete a test with a large specimen, particularly if inert environments are studied in which crack propagation rates are very slow [15]. During this period of exposure to stress and temperature, the initial microstructure may undergo some degradation. It would also be more difficult to maintain a uniform temperature over a large specimen.

A relatively "compact" tapered DCB specimen designed along the lines described by Mostovoy et al. [1] is shown in Fig. 4. In this specimen, approximately constant crack growth rates were observed between crack lengths of 16 mm to 48 mm, providing a useful crack length interval of 32 mm. The variation in crack growth rates over this length was estimated to be within ±4%. The stress intensity factor was calculated from analytical relationships described by Fichter [16] and by Gross and Srawley [17]. Specimens were machined with their longitudinal axes in the direction of rolling.

Procedure

Each specimen was precracked at room temperature and testing was subsequently carried out under the following conditions: static load in vacuum, cyclic load in vacuum, static load in air and cyclic load in air. The number of cycles or time under each type of loading was such that each covered about 8-10 mm of ~32 mm constant-K region. Thus, a microstructure could be evaluated under four different conditions with each specimen. The static loading conditions were such that K = 40 MPa $\sqrt{m}$ and cyclic crack growth testing was carried out at ΔK = 40 MPa $\sqrt{m}$ with R = 0.05 and a frequency of 0.1 Hz.

RESULTS AND DISCUSSION

Static loading

The data shown in Table 2 indicate that the "Conventional" as well as the "Merrick" treatments lead to poor resistance to creep crack growth in air compared to the "Modified" and the "Modified Merrick" treatments. For all microstructures, the mode of fracture in air under static loading was intergranular (Fig. 5).

The poor resistance of the "Conventional" heat treatment has been mentioned by other investigators [18] and this could be attributed to the relatively fine grain size resulting from this treatment. A fine grained material may be expected to undergo a greater amount of grain boundary sliding at elevated temperatures resulting in higher creep rates. Also, it is believed [19] that in an oxygen containing environment, a higher amount of inter-granular embrittlement is possible in a fine grained micro-structure. Such embrittlement is thought to occur through the rapid diffusion of oxygen along the grain boundaries and the consequent oxidation of reactive elements present there.

The low creep crack growth rate in the "Modified" treatment on the other hand, is possibly a result of the relatively coarse grain size produced and the consequently low creep rates. The significant differences in creep crack growth rates between the "Merrick" and the "Modified Merrick" treatments is most likely to be a result of a difference in grain boundary morphology. The irregular grain boundary morphology in the "Modified Merrick" treatment apparently inhibits grain boundary sliding and leads to lower crack growth rates.

In vacuum, creep crack growth rates were generally at least an order of magnitude less than in air testing. The "Modified Merrick" treatment provided the lowest crack growth rates and this once again is attributed to the serrated grain boundary morphology due to this treatment. The fracture mode was primarily intergranular for all cases in vacuum (Fig. 6), but in the case of the "Modified" treatment some transgranular failure could be seen occasionally.

Cyclic loading

Fatigue crack growth rates in air were lowest for the "Modified" heat treatment and highest for the "Conventional" treatment. The rate for the "Modified Merrick" treatment appears to be comparable to that for the "Modified" heat treatment. The fracture mode was transgranular only in the case of the "Modified" heat treatment (Fig. 7) and the superiority of this microstructure under cyclic loading conditions is apparently due to its resistance to intergranular failure (determined in static load tests mentioned earlier). At the low frequency of testing employed, creep effects are likely to contribute significantly to fatigue crack growth and as a result, microstructures which are resistant to intergranular creep cavitation in air can apparently be resistant to fatigue crack growth in air.

In vacuum, the "Modified Merrick" treatment produced the lowest cyclic crack growth rates. The fracture mode for all microstructures in the case of cyclic loading under vacuum was transgranular (Fig. 8).

CONCLUSIONS

The tapered double cantilever beam specimen described in this study has been found suitable for the study of time-dependent fracture in Inconel 718. Based on results obtained from this specimen, the following conclusions have been reached regarding the influence of microstructure upon the creep and fatigue crack growth behaviour of Inconel 718 at 650°C.

1. In laboratory air, the "Modified" heat treatment gives the lowest crack growth rates under static loading ($K = 40$ Mpa $\sqrt{m}$) and under cyclic loading at $\Delta K = 40$ MPa $\sqrt{m}$ and a frequency of 0.1 Hz. The "Modified Merrick" treatment which produces a serrated grain boundary morphology leads to rates which are slightly higher. The "Conventional" as well as the "Merrick" heat treatments result in considerably higher rates of crack extension.

2. In vacuum, the "Modified Merrick" heat treatment resulted in the lowest growth rates both under static and cyclic loading conditions. This effect was clear particularly under static loading conditions indicating that the type of grain boundary serrations due to this treatment are effective in promoting creep crack growth resistance.

REFERENCES

1. H.L. Eiselstein: 'Advances in the Technology of Stainless Steels and Related Alloys', ASTM STP 369, 1965, p. 62.

2. D.F. Paulonis et al: Trans. ASM, 1969, 62, p. 611.

3. W.J. Boesch and H.B. Canada: J. of Metals, 1969, 21, (10),
    p. 34.

4. F.J. Rizzo and J.D. Buzzanel: J. of Metals, 1969, 21, (10),
    p. 24.

5. J.M. Oblak et al.: Met. Trans., 1974, 5A, p. 143.

6. H.F. Merrick: Met. Trans., 1976, 7A, p. 506.

7. R.C. Hall: Trans. ASME, J. of Basic Engineering, 1967, 89
    p. 511.

8. A.K. Koul and G.H. Gessinger, Acta Met., 1983, 31, (7),
    p. 1061.

9. P. Shahinian and K. Sadananda: 'Engineering Aspects of
    Creep', Vol. 2, Paper C299/80; 1980, The Institution of
    Mechanical Engineers, London.

10. A.E. Gemma: Eng. Fracture Mechanics, 1979, 11, p. 763.

11. S. Mostovoy, P.B. Crosley and E.J. Ripling: J. of Materials,
    1967, 2, (3), p. 661.

12. J.P. Gallagher: Eng. Fracture Mechanics, 1971, 3, p. 27.

13. L. Kenyon, G.A. Webster, J.C. Radon and C.E. Turner: 'Creep
    and Fatigue in Elevated Temperature Applications', Vol. 1,
    C156; 1974, The Institution of Mechanical Engineers, London.

14. S. Floreen and R.H. Kane: Fatigue of Eng. Mat. and
    Structures, 1980, 2, p. 401.

15. K. Sadananda and P. Shahinian: Mat. Science and Engineering,
    1980, 43, p. 159.

16. W.B. Fichter: Int. J. of Fracture, 1983, 22, p. 133.

17. J.E. Srawley and B. Gross: Mat. Research and Standards,
    1967, 7, (4), p. 155.

18. K. Sadananda and P. Shahinian: NRL Memo. Rep. 3727, Feb.
    1978, Naval Research Lab., Washington, D.C.

19. J.P. Pedron and A. Pineau: Mat. Science and Engineering,
    1982, 56, p. 143.

Table 1 Heat Treatments for Inconel 718

Heat Treatment Name	Schedule
"Conventional"	$955°C/1h/AC + 718°C/8h \xrightarrow[1°C/min]{FC} 621°C$, hold for a total aging time of 18h, AC
"Modified"	$1093°C/1h \xrightarrow[1°C/min]{FC} 718°C/4h \xrightarrow[1°C/min]{FC} 621°C/16h/AC$
"Merrick"	$1066°C/1h + 843°C/4h + 718°C/8h \xrightarrow[1°C/min]{FC} 621°C/10h/AC$
"Modified Merrick"	$1066°C/1h \xrightarrow[7°C/min]{FC} 843°C/4h + 718°C/8h \xrightarrow[1°C/min]{FC} 621°C/10h/AC$

Table 2 Crack Growth Rates in Inconel 718 at 650°C, in Air and Vacuum for Different Heat Treatment Conditions

Static Loading ($K = 40$ MPa $\sqrt{m}$

Crack Growth Rates (mm/h)

Heat Treatment Name	Air	Vacuum
Conventional	2.67	0.145
Modified	1.42	.065
Merrick	3.00	.049
Modified Merrick	1.5	.0185

Cyclic Loading ($\Delta K = 40$ MPa $\sqrt{m}$, $R = 0.05$, $f = 0.1$ Hz)

Crack Growth Rates (mm/cycle)

Heat Treatment Name	Air	Vacuum
Conventional	3×10^{-3}	5×10^{-4}
Modified	1.4×10^{-3}	7×10^{-4}
Merrick	2.7×10^{-3}	7.7×10^{-4}
Modified Merrick	2.0×10^{-3}	4×10^{-4}

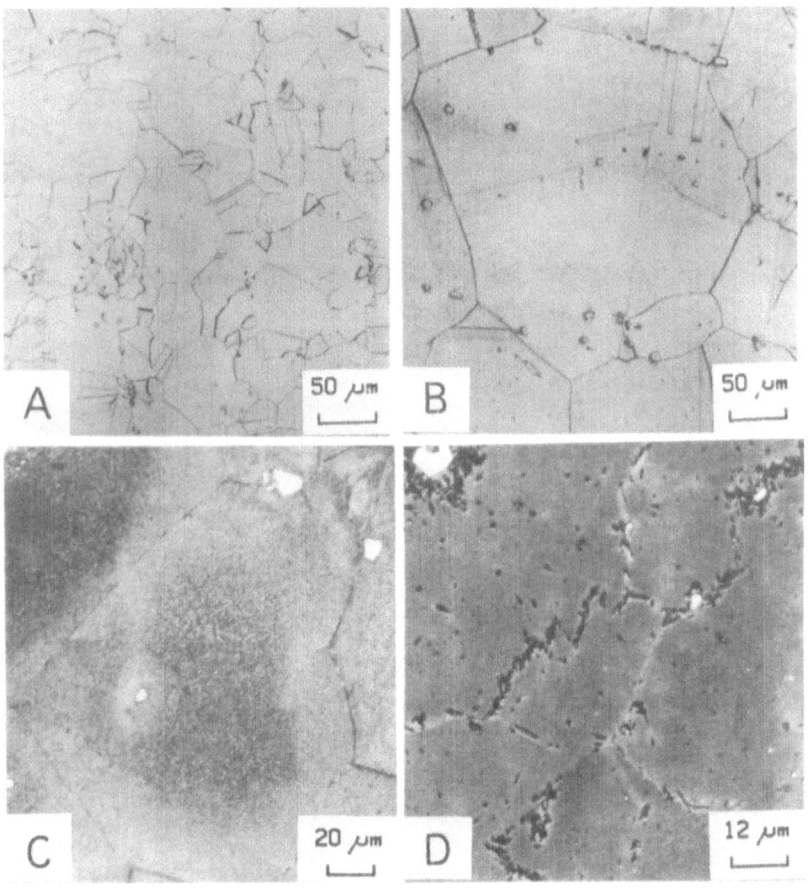

Figure 1   Microstructures of Inconel 718 According to Various Heat
Treatments

    A. "Conventional"   B. "Modified"

    C. "Merrick"   D. "Modified Merrick"

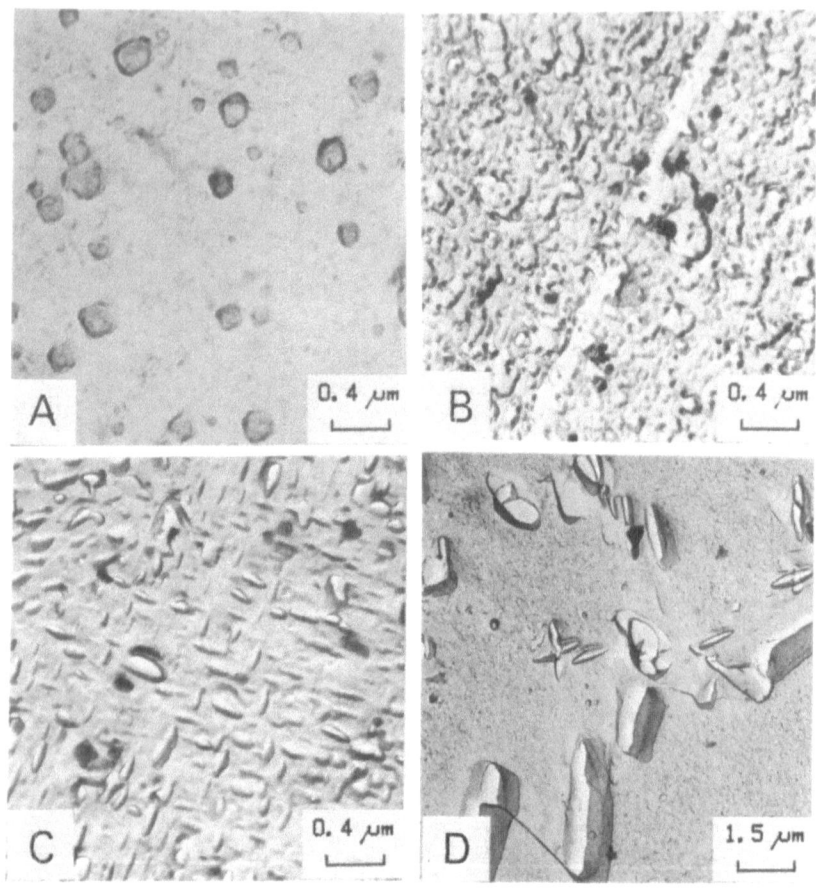

Figure 2 Influence of Heat Treatment on Precipitate Morphologies in
Inconel 718

A. "Conventional   B. "Modified"

C. "Merrick"   C. "Modified Merrick"

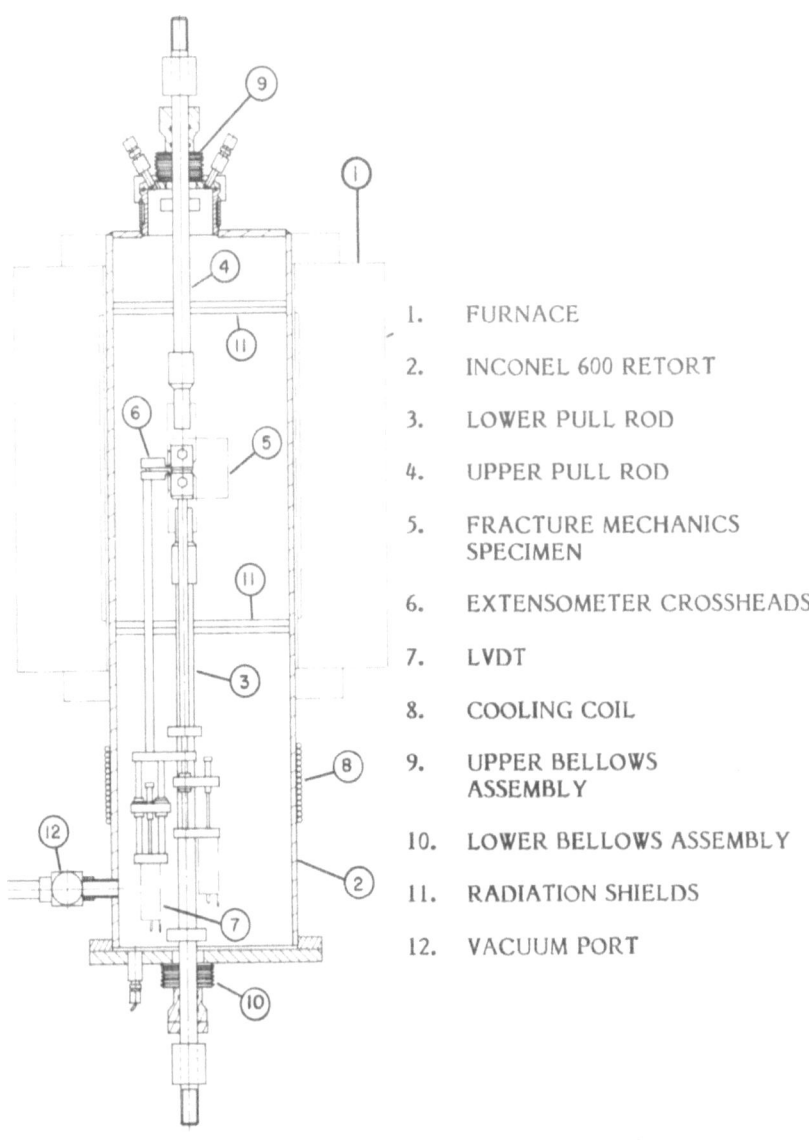

1. FURNACE

2. INCONEL 600 RETORT

3. LOWER PULL ROD

4. UPPER PULL ROD

5. FRACTURE MECHANICS
    SPECIMEN

6. EXTENSOMETER CROSSHEADS

7. LVDT

8. COOLING COIL

9. UPPER BELLOWS
    ASSEMBLY

10. LOWER BELLOWS ASSEMBLY

11. RADIATION SHIELDS

12. VACUUM PORT

Figure 3   Test Assembly for Elevated Temperature Crack growth Rate
            Measurements

256

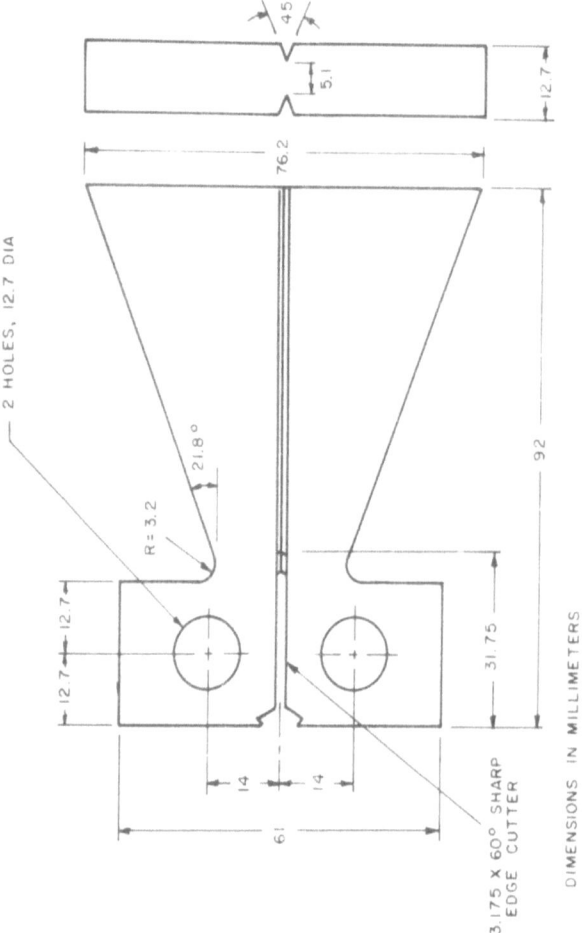

Figure 4   Tapered Double Cantilever Beam Specimen for Constant-K

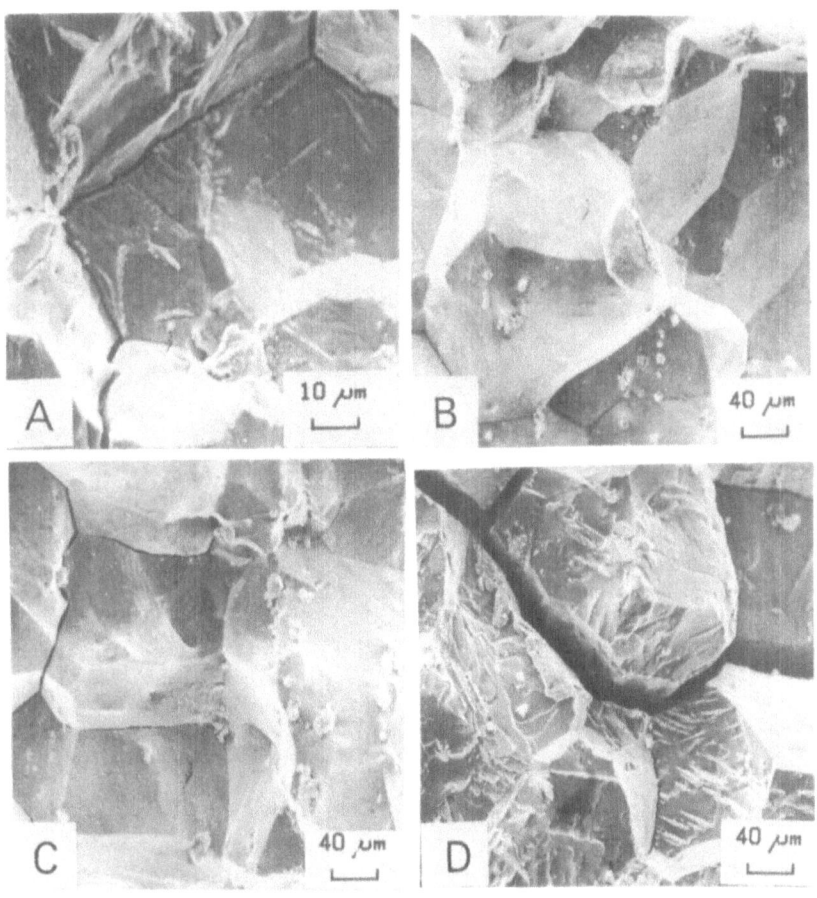

Figure 5   Fracture Modes for Static Loading in Air (K = 40 MPa √m)

A. "Conventional"   B. "Modified"

C. "Merrick"   D. "Modified Merrick"

258

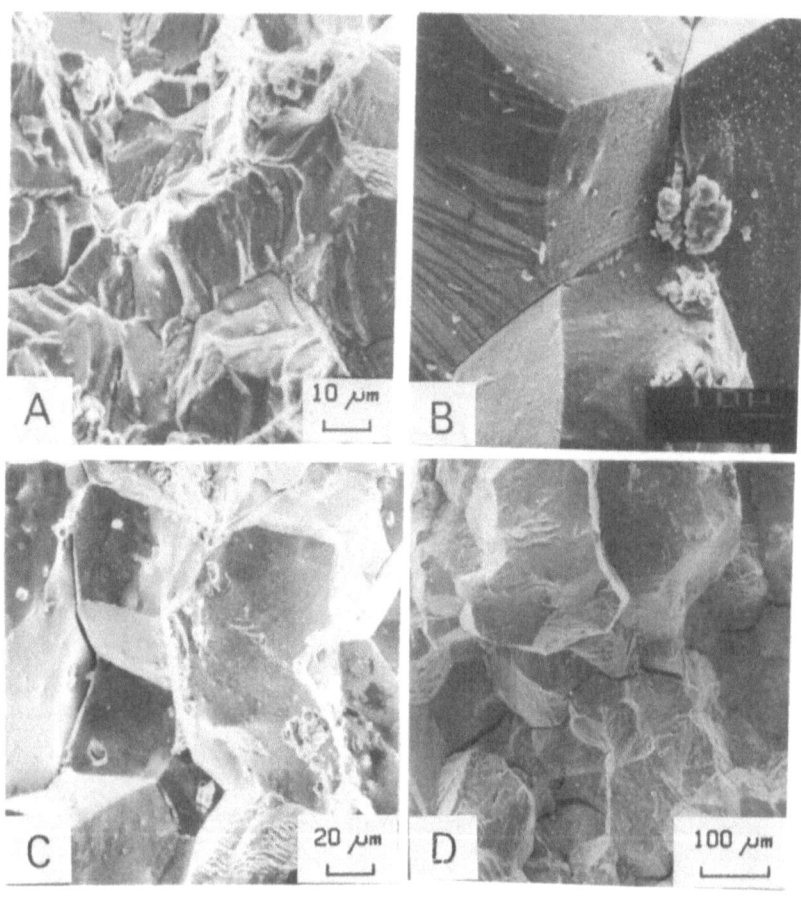

Figure 6   Fracture Modes for Static Loading in Vacuum (K = 40 MPa√m̄)

A. "Conventional"   B. "Modified"

C. "Merrick"   D. "Modified Merrick"

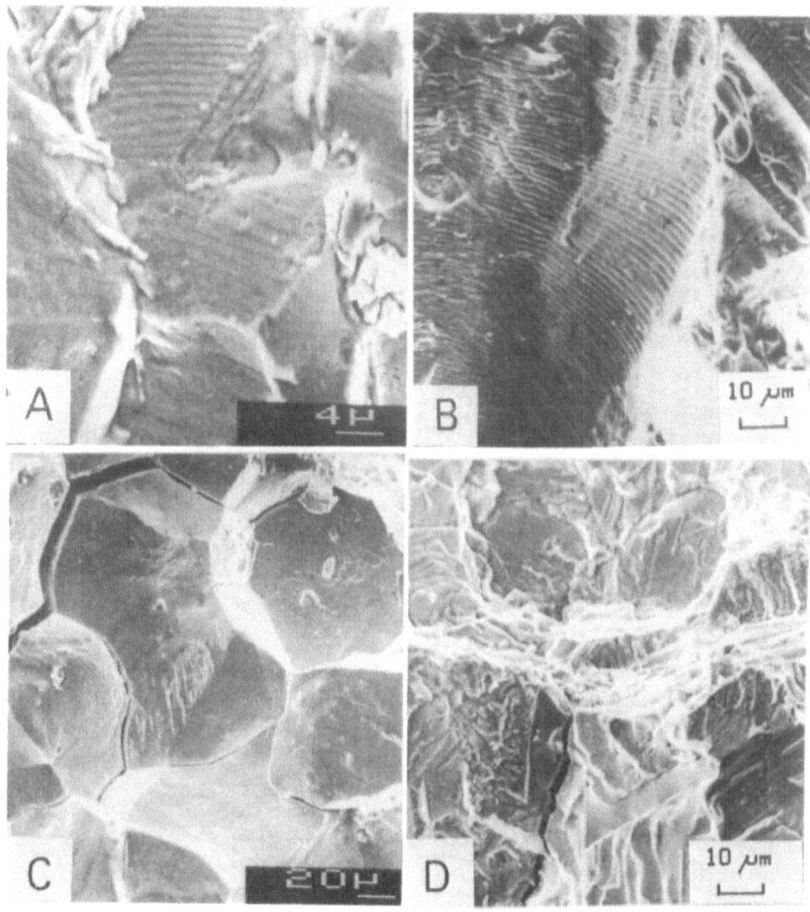

Figure 7   Fracture Modes for Cyclic Loading in Air (ΔK = 40 MPa√m̄,
R = 0.05, f = 0.1 Hz)

A. "Conventional"   B. "Modified"

C. "Merrick"   D. "Modified Merrick"

260

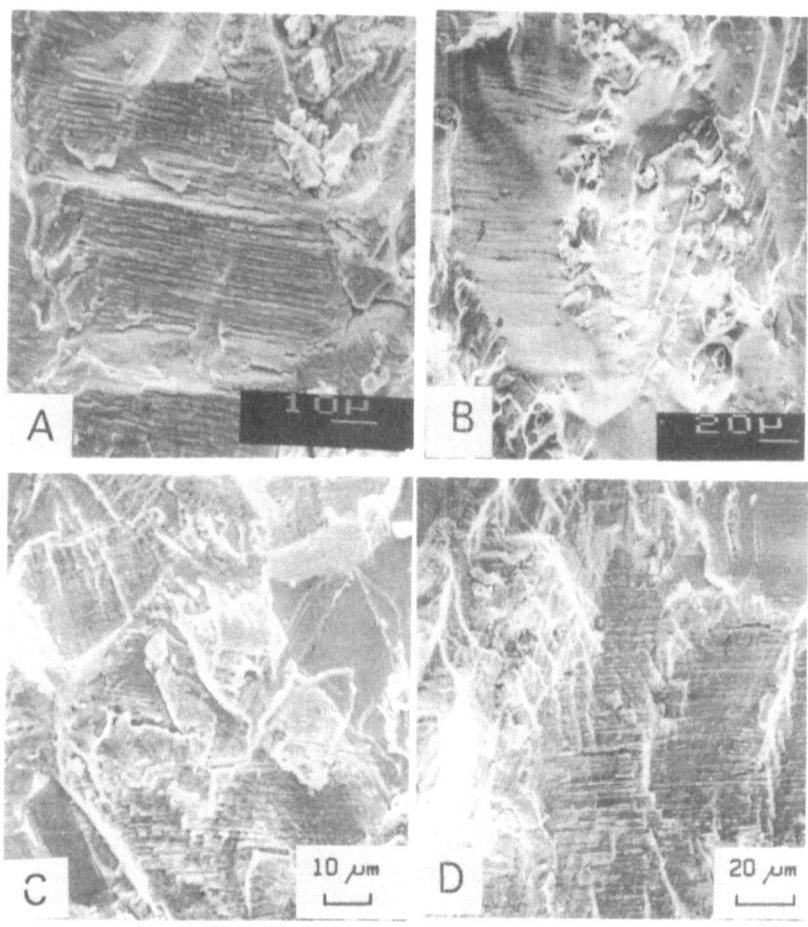

Figure 8   Fracture Modes for Cyclic Loading in Vacuum
($\Delta K = 40$ MPa√m̄, R = 0.05, f = 0.1 Hz)

A. "Conventional"   B. "Modified"

C. "Merrick"   D. "Modified Merrick"

ON THE EMBRITTLEMENT OF NIOBIUM BY OXYGEN

P. G. Watson* and R. E. Reed-Hill

Department of Materials Science and Engineering, University of
Florida, Gainesville, Florida, 32611. *Now at IBM Corporation,
Lexington, Kentucky, 40511.

ABSTRACT

Oxygen strongly reduces the tensile ductility of niobium
between 400 and 1000 K. Because two basically different brittle
fracture modes and ductile rupture (microvoid coallescence) are
competitive in this range the overall fracture process can be
complicated and lead to mixed fracture morphologies. Oxygen
concentration, strain rate and temperature are important parameters. The brittle modes are cleavage which requires a high
stress level and Troiano slow strain rate embrittlement fracture
involving oxygen diffusion and thus favored by slower strain rates
and higher temperatures. Dynamic strain aging is important at the
lower end of the embrittlement range where it increases the work
hardening and raises the stress to the cleavage fracture stress.

INTRODUCTION

It is well known that niobium containing oxygen is subject to
brittle cleavage failure at low temperatures. However, brittle
fracture due to oxygen is also observed at more elevated
temperatures. The first to study the high temperature embrittlement
was Donoso (1) who tested 0.75 mm diameter wire tensile specimens
uniformly doped with oxygen between 77 and 1100 K. Figure 1, from
his results (1), shows reduction in area versus temperature curves
for four oxygen concentrations. Note that at the two lowest levels
(0.09 and 0.40 at. %) the R. A. was very high (0.9) and nearly
temperature independent. However, at 0.85 at. % O a well defined
ductility minimum appeared between 600 and 950 K. Increasing the

262

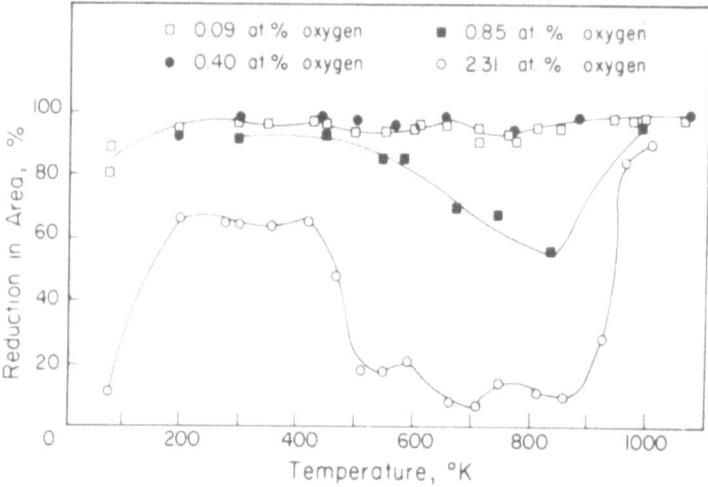

Fig. 1 Donoso's (1) Nb-O ductility curves.

oxygen to 2.31 at. % caused this minimum to be still more
pronounced with the R. A. falling to about 0.1. Figure 1 is
important in that it shows the high temperature oxygen
embrittlement increases in severity with increasing oxygen
concentraton. Donoso also proved that the embrittlement depended
on the strain rate increasing with decreasing strain rate.

SLOW STRAIN RATE EMBRITTLEMENT

An embrittlement which increases in severity and scope with
increasing interstitial concentration and decreasing strain rate is
also characteristic of the slow strain rate hydrogen embrittlement
of steels. A similar tensile strain rate and composition
dependence due to hydrogen in a SAE 1020 steel, was reported in an
early paper by J. T. Brown and W. M. Baldwin, Jr. (2).

The generally accepted explanation for slow strain rate embrittlement is due to Troiano (3) who based his analysis on data
from notched high strength steel specimens tested in static
fatigue. The Troiano mechanism assumes that, during an incubation
period, hydrogen diffuses to the triaxial stress region ahead of
the notch where once the hydrogen level attains a critical value a
crack nucleates. This crack then opens up to the surface of the
notch. Thus the specimen develops a deeper notch with another
stress concentration ahead of it and a second crack nuclei forms
which in turn expands out to the crack front. This process repeats
itself while the crack front advances in a stepwise fashion until

the cross section becomes so reduced that catastropic final
fracture occurs.

Slow strain rate embrittlement due to hydrogen is not limited
to steels since it has also been observed in niobium, vanadium,
tantalum and nickel (4-9).   Nor, as pointed out by Troiano (3), is
hydrogen the only interstitial element that should be capable of
causing slow strain rate embrittlement.

THE TWO FOLD NATURE OF THE OXYGEN EMBRITTLEMENT IN NIOBIUM

While Donoso's experimental results (1) are consistent with
the development of slow strain rate embrittlement in Nb by O between approximately 400 and 1000 K, later work by P. G. Watson (10)
demonstrated that in this interval not one but two embrittling
mechanisms were operative. These two embrittlements are clearly
implied in his R. A. versus T plots in Fig. 2. obtained with Nb
0.75 at. % O specimens. At the higher strain rates ($8.3 \times 10^{-4}s^{-1}$
and $8.3 \times 10^{-5}s^{-1}$) two well defined ductility minima are
apparent. Decreasing the strain rate to $8.3 \times 10^{-6}s^{-1}$, however,
caused both minima to increase in size and overlap so that only one
large minimum appeared to exist.

Raising the oxygen concentration from 0.75 to 1.6 at. % O
increased the embrittlement so that only a single ductility minimum w.
observed over four orders of magnitude of strain rate as shown in Fig

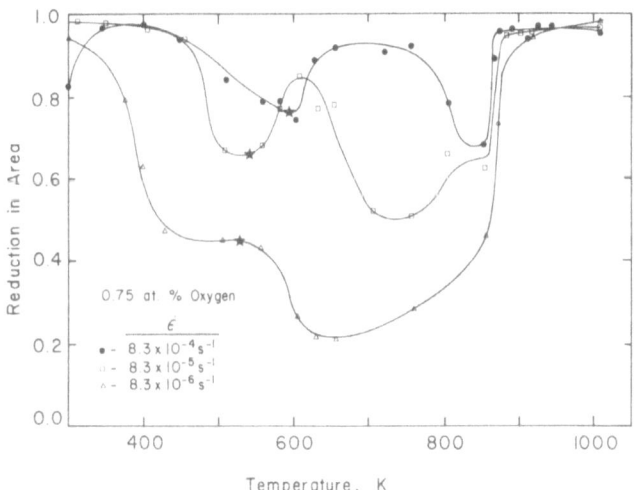

Fig. 2 Watson's (10) Nb-0.75 at. % O ductility curves.

264

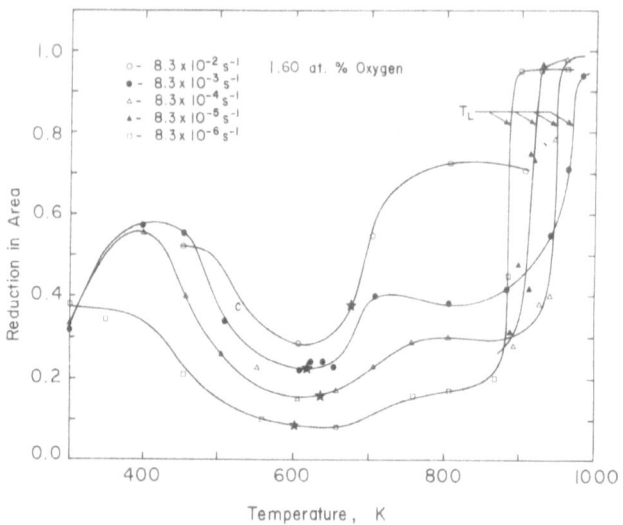

Fig. 3 Watson's (10) Nb-1.6 at. % 0 ductility curves.

It will be demonstrated that the high and low temperature
ductility minima in Fig. 2 correspond to different types of
fracture. However, where these minima overlap the fractures tend
to involve both modes, and the fracture surfaces have a mixed
morphology. It is only near the extremes of the larger combined
ductility minima that fractures tend to exhibit morphologies truly
characteristic of the elementary modes.

THE LOWER TEMPERATURE EMBRITTLEMENT

At the low temperature end of a combined ductility minimum
fracture occurs under very high stresses as can be deduced from
Fig. 4 where the ultimate (true) stress is plotted against the
absolute temperature for 1.6 at. % 0 specimens tested at three
strain rates. Although fractures that started at the surface were
observed above roughly 600 to 650 K, in this low temperature range
surface nucleated cracks were not observed. Thus, below the
asterisks on the ductility curves internal nucleation of fractures
probably occured. Finally the fracture morphology at all strain
rates within this low temperature range was consistent with failure by
cleavage with some brittle intergranular fracture. One is thus led to
the conclusion that here fracture is similar to the subambient brittle
fracture observed in niobium specimens charged with oxygen.

Cleavage fracture usually requires a very high stress level.

Although the critical stress for cleavage may depend on the oxygen concentration, at least at large oxygen concentrations, it tends to be relatively independent of test temperature and the strain required to attain the critical stress. With regard to the stress level required for cleavage of niobium a literature search showed that a specimen of Houssin et al. (11) containing 1.9×10^{-3} at. % O failed at 970 MPa at 77 K by cleavage after 2% strain. Another of Anuchkin and co-investigators (12) with 1.6 at. % O failed brittlely at 77 K just after yielding at 1060 MPa. These data imply that a true stress of about 1000 MPa is probably needed for cleavage in niobium. There are several factors, acting alone or in combination, that can raise the flow stress to this stress. These include increasing the oxygen concentration, lowering the test temperature and increasing the work hardening given to the specimen.

Attention is now called to the work hardening peak that appears in the ultimate true stress versus temperature curves in Fig. 4. This peak is a result of dynamic strain aging due to oxygen. Note that at this peak the stress level rises above 1000 MPa so that cleavage failure should be expected.

Engineering stress strain curves of 1.6 at. % O specimens deformed at the slowest strain rate, $8.3 \times 10^{-6} s^{-1}$ are shown in Fig. 5. R. A. data for these specimens are contained in Fig. 3 and a plot showing the temperature dependence of both the fracture and the ultimate true stresses appears in Fig. 6. As may be seen in Fig. 5, specimens deformed at temperatures below the ultimate stress peak at 450 K necked before fracturing. However, by Fig. 6, the necking strain raised the true stress at the necks above 1100 MPa so that eatastropic brittle fracture occurred. According to Fig. 3 these specimens failed with a R. A. of about 0.37.

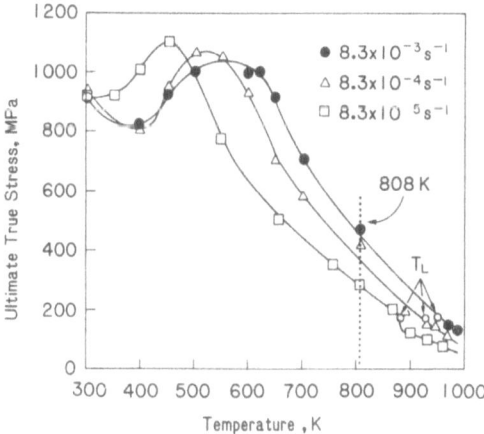

Fig. 4 1.6 at. % O ultimate (true) stress versus T curves.

266

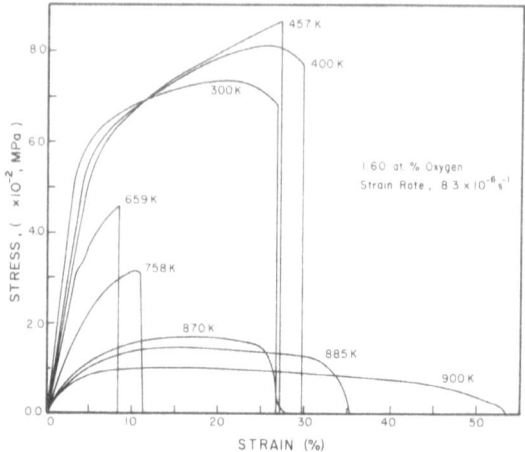

Fig. 5 1.6 at. % 0 stress strain (Engr.) curves. $\dot{\varepsilon} = 8.3 \times 10^{-6} s^{-1}$.

The stress strain curve in Fig. 5 of a specimen deformed at the ultimate stress maximum, 457 K, is significant because it fractured before the start of necking at a true stress level slightly over 1100 MPa as can be deduced from Fig. 6. As a result Fig. 3 shows a drop in R. A. to about 0.21.

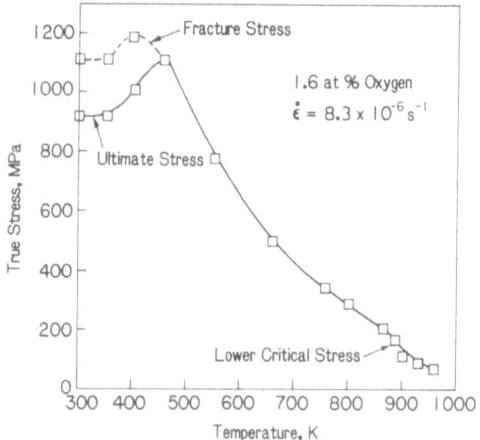

Fig. 6 1.6 at. % 0 ultimate and fracture (true) stresses versus T.
$\dot{\varepsilon} = 8.3 \times 10^{-6} s^{-1}$.

THE HIGHER TEMPERATURE EMBRITTLEMENT

Each ductility curve in Fig. 3 has an asterisk on it marking the lowest temperature at which surface crack nucleation was observed. Surface cracks were found between these temperatures and those marking the sharp ductility increase at higher temperatures.

Figure 4 shows that the ultimate stress decreases rapidly above the work hardening peak where surface crack nucleation occurs. Near the work hardening peak, where the stress level is still high, it is conceivable that the nucleation of a surface crack could precipitate a catastrophic brittle cleavage fracture. However, with increasing test temperature and decreasing stress level surface nucleated cracks should be able to grow larger and larger before the stress rises, due to work hardening, to where rapid catastropic fracture occurs. The fracture surfaces of specimens tested above the work hardening peak show such a mixed character.

Near the high temperature end of a combined ductility minimum, small surface cracks first appear at small strains. An example of showing cracks on a tantalum 0.3 at. % O specimen deformed only to 1.5% strain was given earlier (13). Similar surface cracks on steel specimens charged with hydrogen have been reported. For example, Johnston, Morlet and Troiano (14) show a crack observed on straining a steel specimen 1.5 % at 300 K. With increasing deformation at a low strain rate these small surface cracks grow inwardly across the specimen cross section. The largest cracks ultimately result in the final rupture of the specimen. The growth of these fractures never becomes catastropic and the load of the stress strain curve tends to fall continuously to zero. For example see the 870 K stress strain curve in Fig. 6. Another basic characteristic of the high temperature fractures is that the fracture surfaces tend to show a serrated or stepped morphology. A low magnification SEM photograph of such a surface on a 1.6 at. % O specimen deformed at 884 K at $\dot{\epsilon}$ of 8.3 x $10^{-6}s^{-1}$ was shown earlier (13). A higher magnification SEM micrographs of this fracture surface is shown in Fig. 7. Although this picture implies a fracture process involving intense local plastic deformation the fracture morphology is consistent with the stepwise advance of the crack front in Troiano's slow strain rate embrittlement mechanism.

EFFECT OF STRAIN RATE ON HIGH TEMPERATURE FRACTURE MORPHOLOGY

The slow strain rate Troiano fracture depends on the diffusion of solute over relatively large atomic distances so that a finite time is required, at a given temperature, for a crack to nucleate and additional time is required for fracture to be completed. Thus the Troiano form of fracture is favored by high temperatures and slow strain rates. At 808°K 1.6 at. % O specimens fail by slow

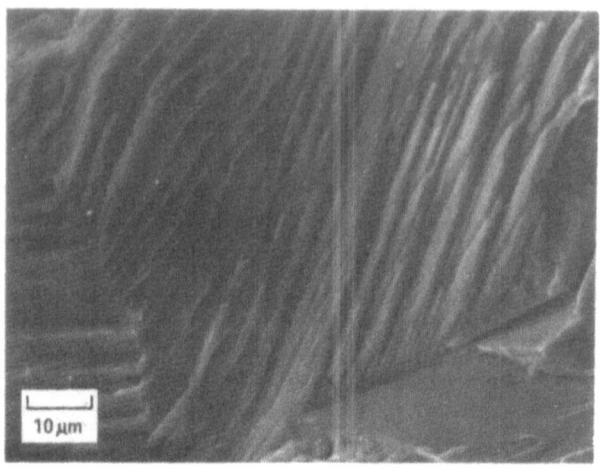

Fig. 7 1.6 at. % 0 fracture surface. 884 K and $\dot{\varepsilon} = 3 \times 10^{-6} \mathrm{s}^{-1}$.

strain rate embrittlement at $\dot{\varepsilon} = 8.3 \times 10^{-6} \mathrm{s}^{-1}$, as shown in Fig.
8a. Increasing the strain rate at this temperature does not
prevent surface cracks from nucleating but reduces the time needed
for them to grow. As a result the stress level rises with
increasing $\dot{\varepsilon}$ as shown by the 808 K in data Fig. 4. However, the
ultimate true stress, within the $\dot{\varepsilon}$ range investigated at 808 K,
falls short of that needed for brittle cleavage fracture so that
cleavage is not competitive at 808 K with the slow strain rate type
of failure. On the other hand, ductile microvoid coalescence
rupture is, and with increasing $\dot{\varepsilon}$ this latter type of failure
becomes more and more significant. This variation in fracture
morphology with strain rate is evident in Fig. 8 where the three
SEM micrographs show specimens deformed to failure at 808 K at
three strain rates. Note that at all three strain rates surface
cracks formed. However, while with increasing $\dot{\varepsilon}$ the number of
cracks increased the depth to which they grew decreased. At the
slowest $\dot{\varepsilon}$ the fracture surface is basically of the brittle Troiano
type. On the other hand with increasing $\dot{\varepsilon}$ the fracture in the interior became more and more ductile. The ductility data in Fig. 3
are consistent with these observations with the R. A. at 808 K
rising from about 0.2 to 0.7 as the $\dot{\varepsilon}$ was increased from 8.3×10^{-6}
to $8.3 \times 10^{-2} \mathrm{s}^{-1}$.

THE LOWER CRITICAL STRESS

     A significant conclusion from the static fatigue tests of
Troiano and co-workers is that there exists a lower critical stress
below which slow strain rate embrittlement does not occur. Tensile

test data also support the existence of a lower critical stress
since at a given ε, if the overall stress level (as measured by the
ultimate stress), in tensile tests made at various temperature,
falls below a critical value, Troiano fracture does not occur and
the ductility sharply increases. Note that the T_L temperatures in
Fig. 3, marking the return to high temperature ductility, define a
stress constant within the experimental error in Fig. 4.

The efforts of P. G. Watson were supported by the Department
of Energy under Contract Number EY-76-5-05-3262.

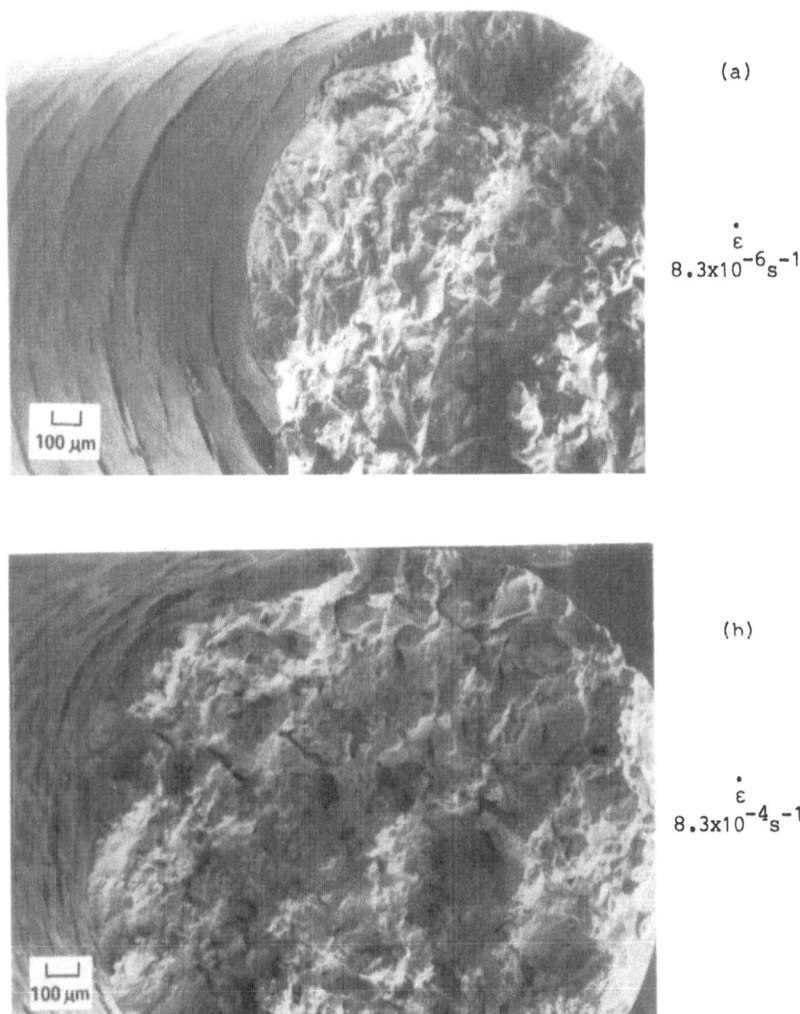

(a)

$\dot{\varepsilon}$
$8.3\text{x}10^{-6}\text{s}^{-1}$

(b)

$\dot{\varepsilon}$
$8.3\text{x}10^{-4}\text{s}^{-1}$

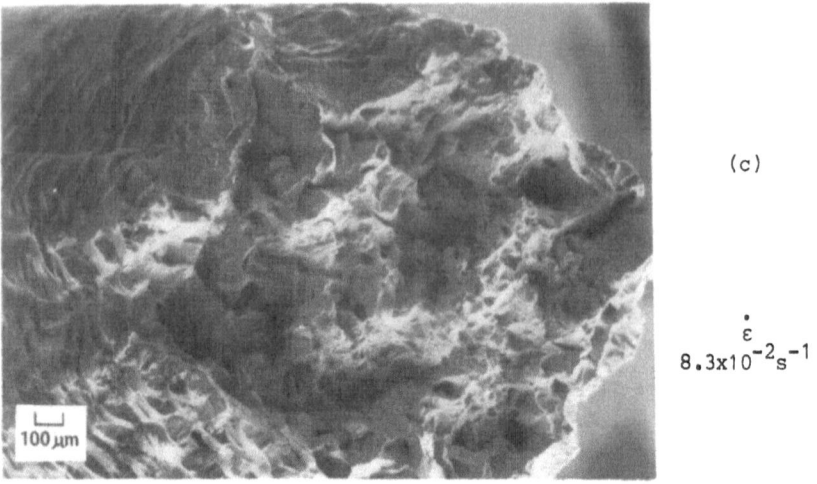

(c)

$\dot{\varepsilon}$
$8.3 \times 10^{-2} s^{-1}$

Fig. 8 1.6 at. % O specimens fractured at three strain rates. 808K.

1. J. R. Donoso and R. E. Reed-Hill. Met. Trans. A 7A(1976)961.
2. J. T. Brown and W. M. Baldwin, Jr., Trans. AIME 200(1954)298.
3. A. R. Troiano, Trans. ASM 52(1960)54.
4. T. W. Wood and R. D. Daniels, TMS AIME 233(1965)898.
5. D. Hardie and P. McIntyre, Met. Tran. 4(1973)1247.
6. A. L. Eustice and O. N. Carlson, Trans. AIME 221(1961)238.
7. D. H. Sherman, C. V. Owen and T. E. Scott, TMS AIME 242(1968)1775·
8. A. G. Imqram, E. S. Bartlett and H. R. Ogden, TMS AIME
    227(1963)131.
9. A. H. Windle and G. C. Smith, Met. Sci. J. 4(1970)136.
10. P. G. Watson, University of Florida, MS Thesis, 1979
11. B. Houssin, J. F. Fries, C. Cizeron and P. Lacombe, Rev. Phys.
    Appl. 5(1970)467.
12. A. M. Anuchkin, A. K. Volkov, V. M. Goritskiy, I. N. Kidin, T.
    Rozhnova, and M. A. Shtremel, Phys. Met. Metall. 34(1972)134.
13. R. E. Reed-Hill, "Hydrogen Effects in Metals," I. M. Bernstein
    and A. W. Thompson, Eds., pg. 873, TMS AIME, New York, NY, 1981.
14. J. G. Morlet, H. H. Johnson and A. R. Troiano, J.I.S.I.
    189(1958)37.

CYCLIC CRACK GROWTH UNDER REPEATED

ROLLING CONTACT

H. Yoshimura, C. A. Rubin and G. T. Hahn

Department of Mechanical and Materials Engineering
Vanderbilt University, Nashville, TN 37235

ABSTRACT

This paper exploits a new technique for installing cracks below the rim of a cylinder to measure the cyclic crack growth produced by rolling contact in 7075-T6 aluminum alloy and AISI-4140 steel. Measurements of crack initiation are also described. The measurements of growth are compared with fracture mechanics predictions based on calculated values of ΔK_{II}, the Mode II crack driving force, and the remotely measured ΔK_{II}-da/dN relation. Various elements of the problem such as: (i) the threshold driving force and flaw size; (ii) crack face friction; (iii) roughness and crack branching; and (iv) the predictability of crack growth are examined.

KEY WORDS

Rolling contact; rolling contact fatigue; crack initiation; cyclic crack growth; LEFM; Mode II cracking; ΔK_{II}-crack driving force; ΔK_{II}-threshold, 7075 aluminum alloy; AISI 4140 steel

INTRODUCTION

The lives of rolling components are foreshortened when cracks in the rim grow large enough under the action of cyclic contact stresses to cause fragmentation of the rim (1-7), e.g., spalling of bearings (1,3), or shelling of rail wheels (2,5). Fleming and Suh (8) were among the first to apply fracture mechanics to the rim cracking problem. The predominantly ΔK_{II}-driving force for Mode II cyclic crack growth has been evaluated for a limited number of crack geometries (9-11). The contribution of friction at the crack faces has been discussed by Rosenfield (13). Finally some measurements of the Mode II cyclic crack growth rate, da/dN, remote from a rim, have been reported (14-17). However, there has been little progress in predicting crack growth under rolling contact, or in defining relations among contact pressure, flaw geometry and rim life. This is partly because of the difficulty of installing well-characterized cracks in the rim of a laboratory specimen.

This paper presents systematic measurements of crack initiation and cyclic crack growth from subsurface defects in rims subjected to 2-dimensional, repeated rolling contact. Two recent analyses of the Mode II, ΔK-crack driving force (18-19) for rolling contact with crack face friction, together with estimates of the da/dN-ΔK_{II} cyclic growth resistance, are used to interpret the measurements. The results shed light on the predictability of rim life and the threshold flaw size.

EXPERIMENTAL MATERIALS AND PROCEDURES

This study was carried out on the 7075-T6 aluminum alloy, a material with a relatively low fatigue crack growth resistance, and on quenched and tempered AISI 4140 steel with a relatively high crack growth resistance. The heat treatments and tensile properties are described in Table 1; the da/dN-ΔK properties are illustrated in Figure 1. Figure 1(c) shows the cyclic growth properties of 300 M steel tempered at 650°C which is similar to the AISI 4140 steel. The test specimens were 50 mm-outside diameter by 35 mm inside diameter by 6 mm-wide cylindrical disks. Crack-like defects, 2a = 0.3 mm and 0.55 mm and inclined at θ = 0°, were installed below the rim of the disks by the hole collapse technique (20) which is illustrated schematically in Figure 2. The disks were heat treated after the defects were installed to remove the effects of the cold work introduced by the indentation. Crack initiation from 0.35 mm- and 0.57 mm-diameter holes drilled through the rim as in Figure 2(a) was also studied. The disk specimens were subjected to repeated rolling contacts under constant loads up to 10 KN in a special machine (20) at speeds of ~ 300 rpm.

Table 1. Heat-treatment Conditions and Tensile Properties

Material	Heat Treatment	Hardness	Yield Strength(MPa)
7075 Aluminum Alloy	T6	HRB - 91	507
4140 Steel	845°C-1 hr. and oil quenched; tempered at 650°C-1 hr.	HRC - 29	889

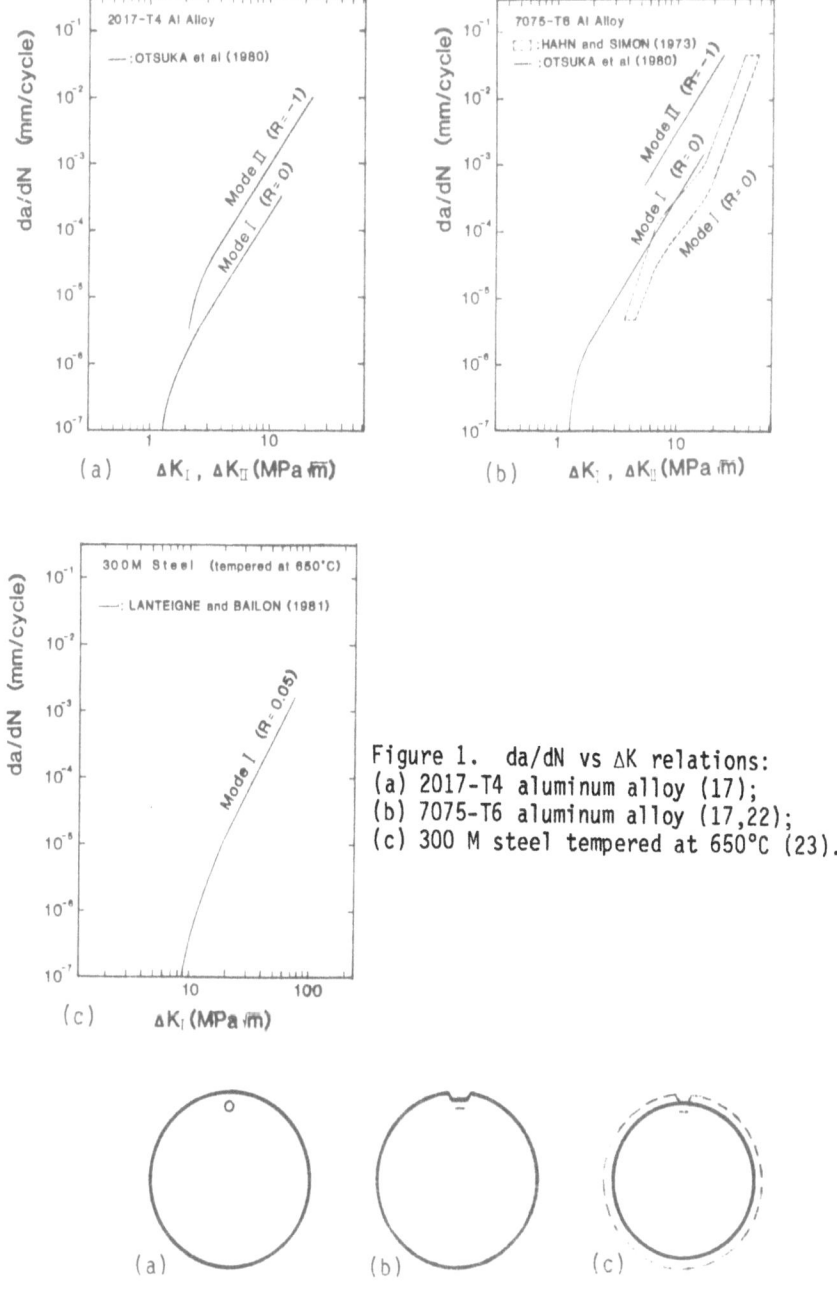

Figure 1.   da/dN vs ΔK relations:
(a) 2017-T4 aluminum alloy (17);
(b) 7075-T6 aluminum alloy (17,22);
(c) 300 M steel tempered at 650°C (23).

Figure 2.   Procedure for inserting a crack-like defect below the
rim of a rolling contact specimen (20): (a) drill small
hole through disk; (b) indent rim to collapse hole, (c)
remove indent by machining and heat treating specimen.

During the rolling operation, the contact surfaces were lubricated
with a lithium-based, molydisulfide high pressure resistant lubricant. The progress of crack initiation and cyclic crack growth
was followed by metallographic sectioning after subjecting the
disks to different numbers of contacts.

RESULTS

7075-T6 Aluminum. Figure 4(a) and (c) illustrates the process
of crack initiation at the drilled holes and the growth of cracks
from this type of defect. The crack nuclei were usually visible after $2 \cdot 10^4$ contacts when the local shear stress range was $\Delta\tau > 160$
MPa. The growth of cracks from the crack-like defects is illustrated in Figure 5 (a, b, and c). For the case of a disk subjected to
10^3 contacts, none of the 4 crack-like defects subjected to $P_0/k = $
3.84 showed obvious signs of growth. However, after $3.2 \cdot 10^3$
repeated contacts, 2 out of 4 cracks subjected to $p_0/k = 4.36$ had
grown. Figures 4(c) and 5(c) both show that the cracks have a wavy
appearance and tend to branch frequently. Comparisons of crack
growth in the forward and backward directions show that, on average,
the cyclic crack growth rate, da/dN of the "trailing" edge is ~ 3x
the value for the "leading" edge [Figure 5(c) and Reference 19].

AISI 4140 Steel. Crack initiation from drilled holes was also
observed in the steel disks when the local shear stress range was $\Delta\tau$
> 330 MPa [see Figure 4(b)], but the crack nuclei failed to grow into
large cracks for the most severe conditions examined: $\Delta\tau = 523$ MPa
and $N = 4.2 \cdot 10^5$ contacts. Similarly, the crack-like defects also
failed to grow under the most severe conditions examined: $\Delta\tau = 420$
MPa and $N = 4.2 \cdot 10^5$ contacts [see Figure 5(d)].

Analysis of Cyclic Crack Growth in 7075-T6 Aluminum. The crack
growth measurements were analyzed in 2 ways. First, the classical
LEFM-approach was used to calculate the number of cycles that produce the observed growth for several assumed values of the crack
face friction. This calculation is based on the da/dN-ΔK_{II} relation for 7075-T6 measured by Otsuka and co-workers(17) and the extrapolation shown in Figure 1(b): da/dN = $A\Delta K_{II}{}^m$ where $A = 3.36 \cdot 10^{-9}$
m = 3 and ΔK_{II}(threshold) ≈ 2 MPa√m. This is combined with expressions for the crack driving force derived by O'Regan and co-workers(18
{for small cracks $\Delta K_{II} = Dp_0 \sqrt{a}$, where $D = f(y/w, \mu^c, p_0/k)$} and
integrated to give the following expression for the number of cycles:

$$N = [An(Dp_0)^m]^{-1}[a_1{}^{-n} - a_2{}^{-n}] \qquad (1)$$

where n = 0.5 m-1 and a_1 and a_2 are the initial and final crack
lengths. The results are given in Table 2, and show that the observed crack extensions are consistent with the calculations for
crack face frictions of $0.6 < \mu_c < 0.8$.

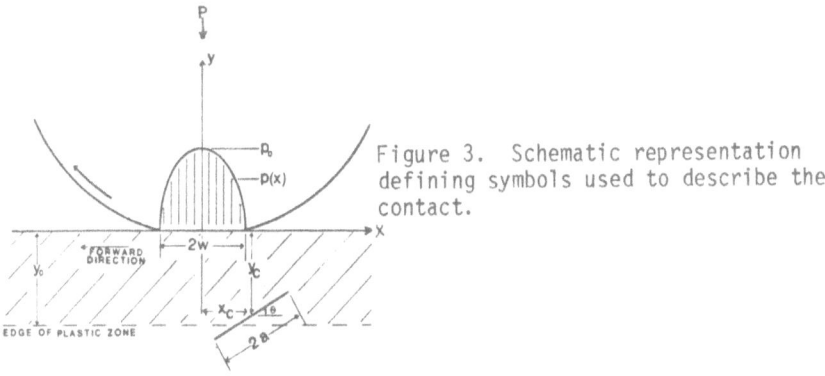

Figure 3. Schematic representation defining symbols used to describe the contact.

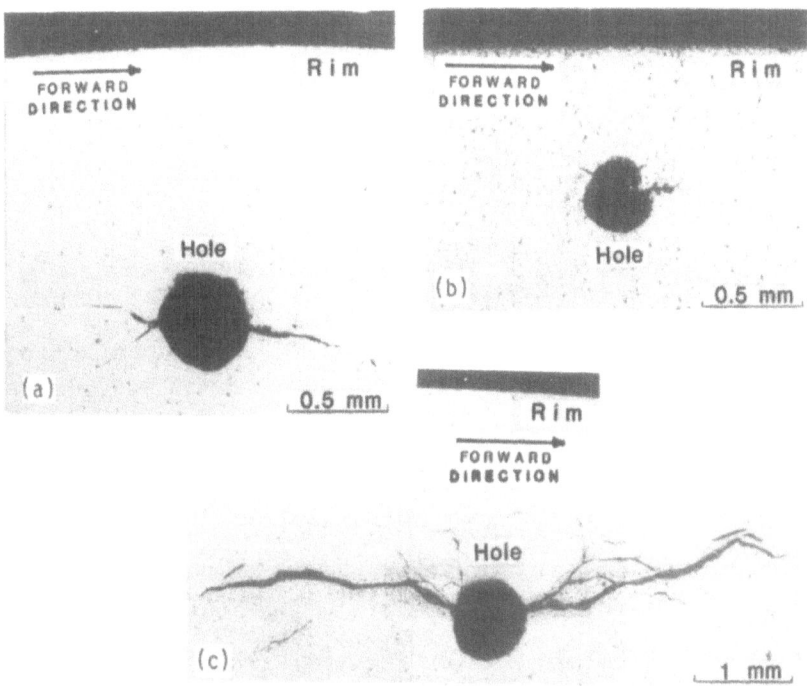

Figure 4. Crack initiation and growth from holes: (a) initiation in 7075-T6 aluminum alloy specimen after N = 1.97·10⁴ at p_0/k = 2.87; (b) initiation in AISI 4140 steel specimen after N = 1.88·10⁴ at p_0/k = 2.65; (c) extensive growth in 7075 aluminum alloy specimen after N = 5 · 10⁴ at p_0/k = 2.88.

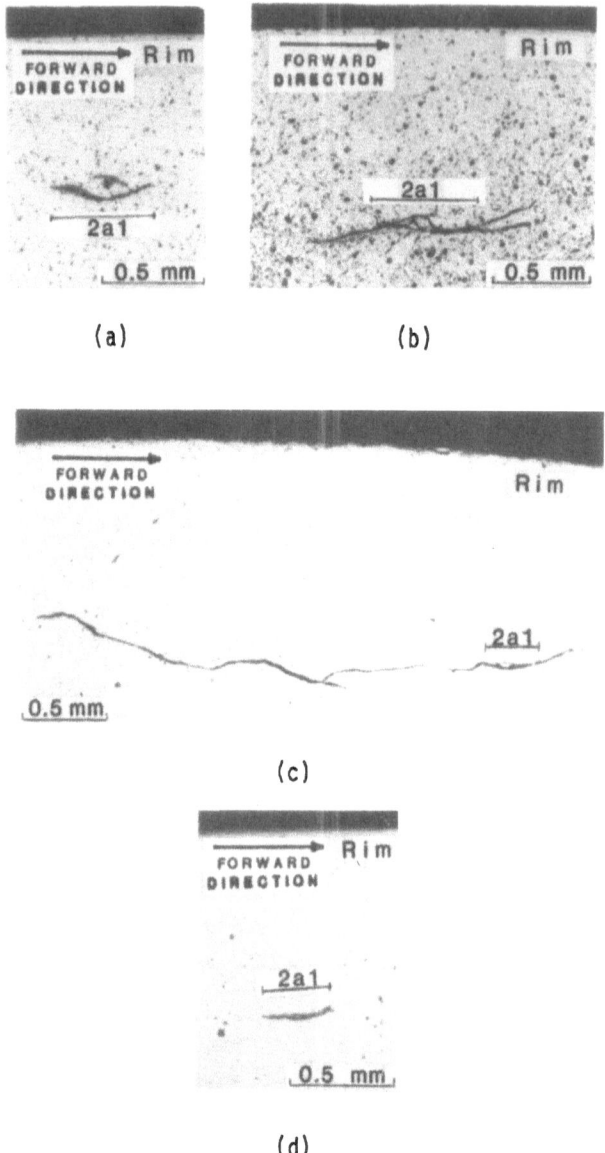

(a)

(b)

(c)

(d)

Figure 5. Crack extension: (a) as installed in 7075-T6 aluminum alloy specimen; (b) growth in 7075-T6 aluminum
alloy specimen after $N = 3.2 \cdot 10^4$ at $p_o/k = 4.36$;
(c) extensive growth in 7075-T6 aluminum alloy spec
men after $N = 5 \cdot 10^4$ at $p_o/k = 2.87$; (d) no
growth in AISI 4140 steel specimen after $N = 4.2 \cdot
10^5$ at $p_o/k = 2.65$.

TABLE 2. Comparison of Experimental Measurements of Subsurface
Cyclic Crack Growth of an Implanted Crack in 7075-T6 Al
Alloy Disks with Fracture Mechanics Calculations

| Experimental Conditions | | | | N, Number of Rolling Contacts | | |
p_0/K	y/W	a_1 (mm)	a_2 (mm)	Actual	Predicted $\mu^c = 0.6$	$\mu^c = 0.8$
2.87	2.13	0.162	1.60	$5 \cdot 10^4$	$2.9 \cdot 10^4$	$1.1 \cdot 10^6$
2.88	1.98	0.275	0.500	$2 \cdot 10^4$	$8.5 \cdot 10^3$	$3.9 \cdot 10^5$
3.28	2.03	0.275	1.77	$7.48 \cdot 10^4$	$1.3 \cdot 10^4$	$5.7 \cdot 10^5$
3.80	2.41	0.275	0.500	$7.2 \cdot 10^4$	$3.6 \cdot 10^3$	$8.4 \cdot 10^4$
3.82	1.42	0.275	0.549	$2.05 \cdot 10^4$	$4.7 \cdot 10^3$	$2.3 \cdot 10^5$
4.36	1.42	0.275	1.549	$3.2 \cdot 10^3$	$6.2 \cdot 10^3$	$3.0 \cdot 10^5$

An alternative analytical approach can be derived by treating
the rough crack profile as a segmented crack: a series of small
cracks separated by regions that are keyed and behave like unbroken
ligaments. For the idealized case of n equal segments of length
$2a_s$ in a large body subjected to a uniform shear stress, τ, the
stress intensity at the ends of the array is equivalent to that produced by a single crack of length $2Fa_s$, where $2 < F < 4$ when $5 < n
< \infty$, and $0.05 < c/a_s < 0.25$, where $2c$ is the ligament width (21).
In this case, the number of contact cycles can be estimated by way
of an average constant growth rate associated with the ΔK_{II}-value
of the ends of the array, i.e., $\Delta K_{II} = Dp_0\sqrt{Fa_s}$.

$$N = [A(Dp_0\sqrt{Fa_s})^m]^{-1}(a_2-a_1) \qquad (2)$$

where a_s corresponds with the "wavelength" of the rough crack. An
analysis of the results illustrates that cyclic lives estimated in
this way for $a_s = 20$ μm agree when $0.2 < \mu^c < 0.4$.

DISCUSSION

The observations of crack initiation and growth during repeated
rolling contact presented here show clearly that Mode II-crack loading can produce cyclic crack growth in compression in the absence
of a Mode I-component. The measurements also support the concept

that fracture mechanics methods can be used to treat the rolling contact spalling process. For example, the measurements of Otsuka et al.(17) in Figure 1a, show that the threshold values of ΔK for cyclic growth in 2017-T4 are nearly the same for Mode I and Mode II crack loading. This correspondence provides a basis for estimating ΔK_{II}, threshold for 7075-T6 aluminum alloy and AISI 4140 steel [see Figure 1(b) and (c)]. A comparison of these threshold values with the ΔK_{II}-values calculated for the defects in the disks for $\mu_c = 0.4$ in Table 3 provides a rationale for the observation of cyclic crack growth in the 7075-T6 disks where $\Delta K_{II} > \Delta K_{II,\text{threshold}}$, and not in the AISI 4140 disks where $\Delta K_{II} < \Delta K_{II,\text{threshold}}$. Further, the Hearle and Johnson fracture mechanics analyses (19) predict that ΔK_{II}-values for the trailing edge exceed those for the leading edge of the crack by 45% when $a/y > 1$. For the case of the 7075-T6 aluminum alloy, where $da/dN \propto (\Delta K)^3$, this 45% prediction translates into a 3-fold extension for the trailing edge compared to the leading edge. This difference is confirmed by the present measurements. The fracture mechanics analysis coupled with $da/dN - \Delta K_{II}$ measurements thus have the potential for defining the effects of contact pressure, initial crack size, location, inclination, etc., on the contact life of a rim. For example, estimates of the threshold crack size (cracks smaller than this would not be expected to grow) for 7075-T6 and AISI 4140 steel for $p_0/k = 4$ ($p_0/\sigma_0 = 2.3$) are listed in Table 3.

The present observations provide circumstantial evidence that crack face friction plays a role in cyclic crack growth under the contact. The fact that cracks initiate at drilled holes in the AISI 4140 steel disks but then fail to grow [see Figure 4(c)] and that crack-like defects also fail to grow [see Figure 5(d)] are signs that cyclic growth is impeded by the rubbing crack faces. This is also confirmed quantitatively by the estimates of the crack driving force for AISI 4140 in Table 3, which fall below the threshold value when a reasonable crack face friction $\mu^c = 0.4$, is invoked.

The treatment of crack face friction is complicated by the rough wavy crack surfaces which may be connected with the crack branching. The roughness manifests itself either as an artificially

Table 3. Summary of Initial ΔK_{II}-Driving Force, and Threshold Crack Length Values for 7075-T6 Al Alloy and AISI 4140 Steel.

Material	$\Delta K_{II}(a = a_i)$, MPa√m		$\Delta K_{II,\text{thresh}}$ MPa√m	$a_{\text{threshold}}$	
	$\mu_c = 0$	$\mu_c = 0.4$		$\mu^c = 0$ μm	$\mu^c = 0.4$ μm
7075-T6	4-16	2-6	~ 2	2	13
AISI 4140	4-15	3-6	~ 9	25	150

high and probably crack length-dependent value of μ^c (see Table 2) when the crack is treated as a smooth continuous planar flaw, or as a "keyed" spot that can be treated like an unbroken ligament. The segmented crack model offered here is consistant with more realistic values of μ^c and has the virtue that it produces a relatively crack length-insensitive crack driving force. It remains to be seen whether this type of model can provide the large difference in the growth rates displayed by the trailing and leading edges.

The development of appropriate fracture mechanics analyses of rolling contact fatigue will benefit from: (i) the definition of the Mode II, da/dN-ΔK curve of the experimental materials, (ii) calculations of the Mode II crack driving force that account for roughness and departures from linearity and (iii) measurements of the cyclic crack growth rate under the contact as a function of crack length. The mechanisms of crack initiation and the effects of the continuing cyclic deformation of the rim must also be addressed.

CONCLUSIONS

1. Crack initiation and cyclic crack growth from subsurface defects proceeds in the rim of 7075-T6 aluminum alloy disks subjected to N = 3.2 $\cdot$ 10^3 to N = 7.5 $\cdot$ 10^4 repeated rolling contacts at peak pressures from p_0/k = 2.4 to p_0/k = 4.4.

2. The cracks, which are produced by a Mode II cyclic crack loading under compression, are rough and branch frequently. The trailing edges of the cracks tend to grow about 3 times as fast as the leading edges.

3. Cracks initiate in AISI 4140 steel disks after N = 8 $\cdot$ 10^3 to N = 4.2 $\cdot$ 10^5 contacts, but these cracks and the installed crack-like defects fail to grow at pressures up to p_0/k = 3.4.

4. The cyclic growth in the aluminum alloy disks and absence of growth in the steel are consistent with fracture mechanics analyses that account for crack face friction and roughness.

ACKNOWLEDGEMENTS

This paper is based upon work supported by the National Science Foundation under Grant No. DMR-8108500. The authors wish acknowledge invaluable contributions made by W. Wright and W. Gentry to the design and production of the rolling contact machine. They wish to thank J. Hightower and R. McReynolds for their assistance in the laboratory and C. Wieger for her work on the manuscript. They are also grateful to K. L. Johnson of Cambridge University for communicating results of his work with A. D. Hearle.

REFERENCES

1. W. D. Syniuta and C. J. Corrow, Wear, 15 (1970), 187-199.
2. G. C. Martin and W. W. Hay, ASME, 72-WA/RT-8, 1972.
3. K. Sugino, K. Miyamoto, and M. Nagumo, Trans. ISIJ, 11 (1971), 9-17.
4. N. P. Suh, Wear, 44 (1977), 1-16.
5. T. Kunikake, S. Nishimura and H. Tagashira, Trans. ISIJ, 10 (1970), 476-489.
6. D. M. Feyredo and C. Prichard, Wear, 49 (1978), 67-68.
7. Tokuda, Nagatuchi, Tsushima and Muro, ASTM STP771, pp. 150-165.
8. J. R. Fleming and N. P. Suh, Wear, 44 (1977) 39-56.
9. L. M. Keer, M. D. Bryant and G. K. Haritos, J. Lubrication Technology, July 1982, Vol. 104, 347-351.
10. L. M. Keer and M. D. Bryant, ASME, 82-Lub-10.
11. C. G. Chipperfield and A. S. Blicblau, (private communication).
12. D. A. Hills and D. W. Ashelby, Eng. Fracture Mech., Vol. 13, 69-78.
13. A. R. Rosenfield, Wear, 61 (1980) 125-132.
14. R. Roberts and J. J. Kibler, Trans. ASME, Series D, 93, 671-680.
15. L. P. Pook, Int. J. Fracture, 13 (1977), 867-869.
16. L. P. Pook and A. F. Greenan, Fracture Mechanics, edited by C. W. Smith, ASTM STP 677, 23-35.
17. A. Otsuka, K. Mori, T. Ohshima and S. Tsuyama, Advances in Fracture Research, edited by D. Francois, et al., Pergamon Press, U.K., 1980.
18. S. O'Regan, G. T. Hahn and C. A. Rubin (submitted to Wear).
19. A. D. Hearle and K. L. Johnson (private communication).
20. H. Yoshimura, C. A. Rubin and G. T. Hahn, Wear, 95 (1984) 29-34.
21. D. P. Rooke and D. J. Cartwright, Compendium of Stress Intensity Factors, Her Majesty's Stationery Office, London, 1976, 140-141.
22. G. T. Hahn and R. Simon, Eng. Fracture Mech., 1973, Vol. 5, 523-540.
23. J. Lanteigne and J. P. Bailon, Metallurgical Trans. A, Vol. 12A, March 1981, 459-466.

APPENDIX

Some of the symbols are defined in Figure 3. Other definitions are given below:

a,threshold: largest crack that does not grow; F: segmented crack length coefficient; k: shear yield strength = $\sigma_0/\sqrt{3}$; $\Delta K = K_{max} - K_{min}$; K_I, K_{II}: Mode I, Mode II stress intensity factors; N: number of rolling contact; P: contact load; p_0: peak contact pressure; σ_0: yield strength; $\Delta\tau = \tau_{max} - \tau_{min}$; τ: Hertzian shear stress; μ^c: crack surface friction coefficient.

THE DEVELOPMENT OF CURVED FRACTURE TOUGHNESS SPECIMENS FOR
PREDICTING CRACK GROWTH IN CANDU REACTOR PRESSURE TUBES

A.C. Wallace

Atomic Energy of Canada Limited
Chalk River Nuclear Laboratories
Chalk River, Ontario, KOJ 1J0 Canada

ABSTRACT

The conventional method for estimating critical crack size in
CANDU reactor pressure tubes has been to burst segments of tube
containing defects. Since burst tests are expensive in terms of
both cost and material, alternative tests were developed using
small fracture toughness specimens cut from flattened sections of
tube. Critical crack length was determined using an elastic-
plastic crack growth resistance curve (R-curve) technique. Due to
the dependence of R-curves on specimen geometry and the difficulties of flattening brittle, irradiated pressure tube, a program
was undertaken to develop small fracture toughness specimens
retaining the original tube curvature. Three types of curved
specimens are discussed with emphasis on the centre cracked type.
The operation of specially designed grips required to test the
specimen are described and the results of a preliminary finite
element analysis of the specimen are discussed.

1. INTRODUCTION

The core of a CANDU nuclear reactor consists of 380-480
cold-worked Zr-2.5 wt% Nb pressure tubes which contain the fuel
and the heavy water coolant. The tubes are nominally 6.3 m long,
with a mean diameter of 107 mm and a 4.1 mm wall thickness, and
operate at an internal pressure of 9.8 MPa and a temperature of
about 550 K. Determination of the critical crack size for unstable

crack propagation in pressure tubes is a major concern since it establishes whether or not leakage can be detected from a crack prior to tube rupture.

Traditionally, critical crack length has been estimated from burst tests on 500 mm sections of pressure tube containing machined flaws (1). It has long been recognized that the collection of critical crack length data from tests on small fracture toughness specimens offers many advantages over the burst-testing technique (2). Small specimen tests are simpler and less expensive to perform than burst tests and require consider-ably less material. Specimen and experimental characteristics are more easily changed, permitting a wider range of variables to be examined, and a more statistically significant analysis of results.

Most of the small specimen fracture toughness tests have been performed using compact tension specimens machined from flattened pieces of pressure tube. However, the flattening operation reduces the wall thickness of the specimen and introduces residual stresses which may affect the results. Moreover, in the case of brittle material, irradiated to a high fluence or containing high hydrogen concentrations, flattening operations can break the specimens. For these reasons, attention has recently turned towards the development of small fracture toughness specimens which retain the curvature of the pressure tube.

2. CANDIDATE SPECIMEN DESIGNS

Three different types of curved specimens are being developed - the centre cracked tension specimen, the curved compact tension specimen and the curved double-torsion specimen.

The curved double torsion specimen (Fig. 1) is a variation of the double-torsion specimen used by ceramicists. Although unusual in appearance, the crack tip experiences mode I, or tensile opening mode loading, and the specimen has the added feature of being a 'constant K' specimen, meaning that within a certain range the stress intensity factor, K_I, is independent of crack length (3). It is expected that the curved double-torsion specimen will behave in the same manner, provided that the specimen is made small enough to minimize any effects of curvature (4). The current design is for a specimen to be one-twelfth of the circumference of the pressure tube, (i.e. dimension T in Figure 1 = 27 mm). This specimen is currently undergoing further development at Ontario Hydro's Research Division.

The curved compact tension specimen (Fig.2), is spark machined from an undeformed section of pressure tube in a single operation, using a 'cookie-cutter' electrode. The small size of

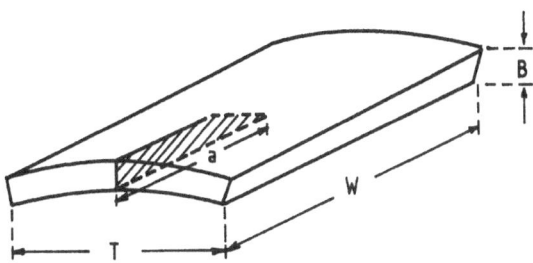

FIGURE 1   Curved double torsion specimen

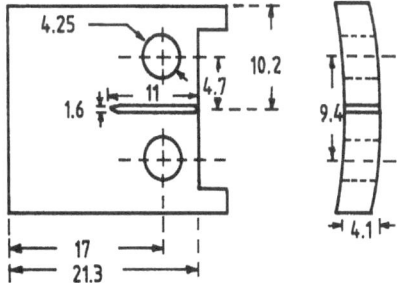

FIGURE 2   Curved compact tension specimen - dimensions in mm

the specimen minimizes the effects of bending in the crack plane
due to the offset loading axis. Tests have shown that identical
results are obtained using flattened and curved specimens from the
same material. Thus this specimen has the advantage that it can
use the analyses already available for the flat compact tension
specimen in both the elastic and elastic-plastic regimes.

Due to the small size of these specimens, the theories and
techniques of elastic-plastic fracture mechanics must be used to
obtain valid predictions of pressure tube failure. Critical crack
length is determined from the point of tangency of an experimentally derived crack-growth resistance curve, based on the
J-integral, and a theoretical crack driving force curve (5).
Comparisons of the resistance curves derived from flat centre
cracked and compact tension specimens have shown that the curves
are geometry dependent. R-curves from centre cracked specimens
are steep and predict critical crack lengths in good agreement
with burst test results while R-curves from compact tension
specimens are less steep and give slightly conservative results
(6). Thus there is interest in developing a small curved specimen
with a centre cracked geometry.

284

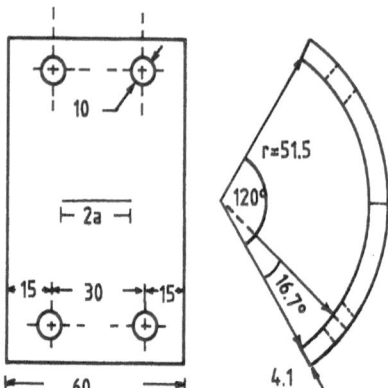

FIGURE 3 Curved centre cracked specimen - dimensions in mm

The curved centre cracked specimen (Fig. 3) is simply one third of the circumference of a section of pressure tube 60 mm long. A starter notch is spark machined in the centre and sharpened by fatigue. The specimen is pin-mounted to specially designed grips through the four large holes seen in Fig. 3. These grips provide the key to testing the curved centre cracked specimen.

3. CURVED SPECIMEN TESTING GRIPS

Fig. 4 shows a side view of the specimen mounted in the grips. The grips consist of two interlocking sections, a, connected to the side plates, b, by pins, c. The specimen, d, is fixed to the grips by four pins through the L-shaped plates at e. When a tensile load is applied from the clevis and pin assemblies, f, the two halves of the grips, a, will pivot slightly about their respective connecting pins, c. Because these pivot points are offset from the centre of curvature of the specimen, the curved surfaces of the grips, which initially follow the inside surface of the specimen exactly, push outwards on the inside of the specimen, constraining it to deform in a manner approximating that of a tube under increasing internal pressure.

The operation of the grips is demonstrated more clearly in Figs. 5 and 6, which show the grips with specimen removed. The curved surfaces of the grips consist of interlocking 'fingers' which are initially closed (Fig. 5) and open up as a load is applied (Fig. 6). This motion increases the radius of curvature of the grip curved surface, and consequently that of the specimen. The increase in radius of curvature is not uniform over the length

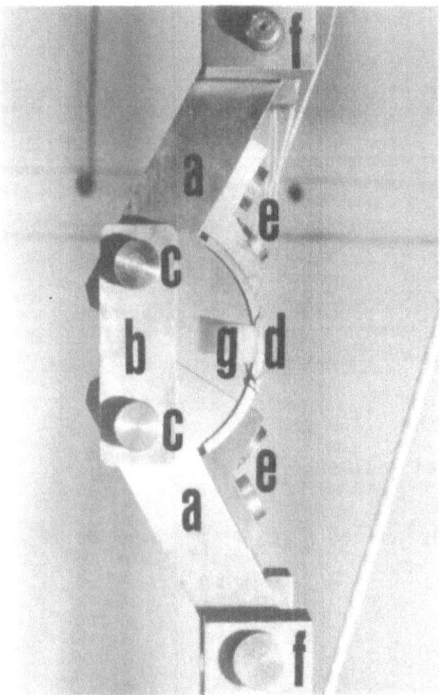

FIGURE 5 Grips in initial closed position.

FIGURE 4 Curved centre cracked specimen mounted in grips.

FIGURE 6 Grips after opening.

of the specimen. However, calculations have shown that the variation amounts to only a few hundredths of a millimetre at the largest expected displacements.

For a given rotation about the pivot points, ('c' in Figure 4), it is possible to calculate both the average radius increase and the strain produced on the inside surface of the specimen. The geometry of the grips was chosen so that the strain calculated from the length increase of the specimen inside surface agreed as closely as possible with the strain calculated from thin-walled pressure vessel theory, assuming a deformed radius based on the calculated average radius increase. The strains determined in this manner are compared in Table 1 for grip half pivot point rotation angles of from 1° to 5°. The two values agree quite well suggesting that the grips constrain the specimen to deform in a manner approximating that of a tube under pressure.

TABLE 1 Comparison of Strain Produced in Specimen by Grips and
Strain Predicted from Theory[1] Based on Grip Average
Radius Increase

Rotation Angle (Degrees)[2]	Average Radius Increase (mm)	Strain on Deformed Surface	Strain Predicted From Theory
1	0.6015	.011677	.011680
2	1.2058	.023415	.023414
3	1.8119	.035192	.035183
4	2.4190	.046981	.046971
5	3.0261	.058789	.058759

1. $\varepsilon_{cir} = \dfrac{R_i - r_i}{r_i}$

r_i = undeformed internal radius

R_i = average deformed internal radius

2. Rotation angle applies to one half of grips - the other half
will also rotate by the same amount.

4. TESTING EXPERIENCE

The variation in radius increase along the specimen during
loading is such that the radius is slightly larger on either side
of the crack plane than it is directly beneath it. There is thus
a tendency for the specimen to 'flatten' slightly in this region,
producing bending stresses in the crack plane. These stresses
have a maximum tensile value on the specimen I.D. and can cause
uneven fatigue precrack fronts such as that shown in Figure 7.
This behaviour may be avoided by installing a narrow strip of
teflon about .08 mm thick underneath the crack. This provides
sufficient support to the specimen to eliminate the bending stress
producing straight precrack fronts such as that shown in Fig. 8.
The teflon also helps to reduce any friction between the grips and
the specimen. Since very little, if any, sliding takes place
during a test and most deformation occurs near the crack plane,
friction is not thought to have much effect. However, very large
contact forces exist between the grips and the specimen, so a
graphite coating is applied to the contact surfaces to encourage
sliding.

Fig. 9 shows the full testing arrangement with the
displacement gauge installed. Elastic bands are used to hold the
gauge in place since it tends to wobble during a test, spoiling
the results. Figure 10 shows the assembly with the displacement
gauge removed. The crack extension gauges used to monitor crack
growth during a test are visible on either side of the notch.

I.D.

FIGURE 7 Uneven fatigue crack
front due to bending – no teflon.

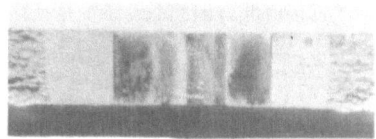

I.D.

FIGURE 8 Straight fatigue
crack front produced by
inserting thin teflon strip
under specimen notch.

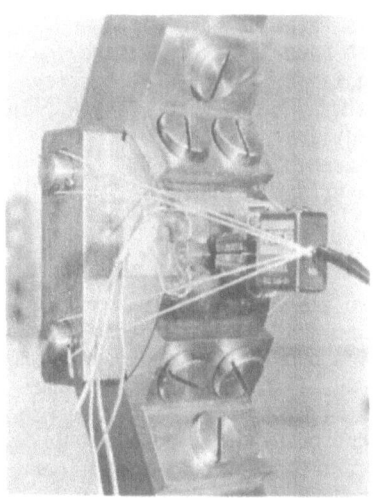

FIGURE 9 Specimen mounted in
grips ready for testing.

FIGURE 10 View of crack
extension gauges for crack
length monitoring.

Only a few tests have been conducted to evaluate the
performance of the grips. The instrumented specimen shown in the
photographs has not yet been tested but the grips have peformed
well. The specimens shown in Figs. 7 and 8 failed in a completely
brittle manner, as expected, with no detectable distortion of the
curvature of the specimen. However, the results of these and
other tests cannot be properly evaluated in the absence of a
detailed stress analysis of the specimen.

As an interim measure, results have been analyzed using the following relation for a flat centre cracked plate in tension (7):

$$K_I = \frac{YP\sqrt{a}}{BW}$$

where Y $= 1.77\ (1-0.1(2a/W) + (2a/W)^2),\ 0 \leq 2a/W \leq 0.6$ (1)
    P = applied load
  2a = crack length
    B = specimen wall thickness
    W = specimen width
   K_I = mode I stress intensity factor

Critical K_I values obtained using this relation were within the expected range, but there was some uncertainty about the validity of the results. An experimental compliance calibration is planned to develop the relation between K and crack length. In the interim a simplified finite element analysis has been performed to determine the load-displacement and load-K_I relationships for the grips and the specimen.

5. FINITE ELEMENT ANALYSIS

The analysis was performed in two stages. First, a simplified two-dimensional analysis was performed to determine the load-displacement response of the grips and an unnotched specimen. The analysis used plane strain, 8-noded, insoparametric quadrilateral elements. The mesh is shown in Figure 11 with the specimen being represented by the four thin, curved elements adjacent to the grip curved surface. Only half of the specimen is modelled. The contact region between the specimen and grip elements is modelled using gap elements which constrain the surfaces to slide over one another, yet permit a normal force to be transmitted. From this analysis, the displacements of the specimen nodes in contact with the grips were determined for given grip rotations.

The second phase of the analysis used a three-dimensional mesh of 20-node brick elements representing one quarter of the specimen, Figure 12. This mesh was used to determine the linear elastic stress intensity factor, K_I, for a given load. The inside diameter nodes of the mesh were constrained to undergo the displacements determined from the previous analysis, applying the appropriate boundary conditions on the 3-D mesh. The load corresponding to a given displacement was determined from the reaction forces at the nodes fixed in the crack plane ahead of the crack tip. The two-dimensional analysis has shown that the connecting bars, b in Fig. 4, do not transfer any of the applied tensile load. This was confirmed with strain gauges.

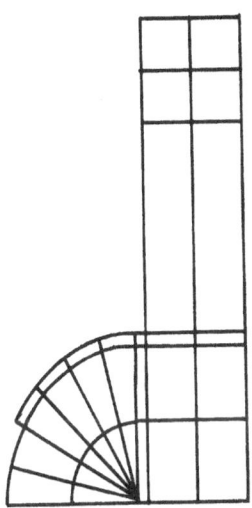

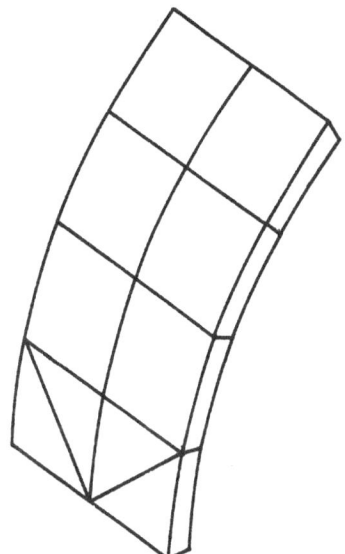

FIGURE 11 Two dimensional finite
element mesh used to determine a
grip-specimen deformation
behavoiour.

FIGURE 12 Three dimensional
finite element mesh used to
determine stress intensity
factor, K_I.

TABLE 2 Comparison of Stress Intensity Factors Calculated by
Finite Elements and Equation (1) (2a/W = 0.5)

Applied Load	Stress Intensity Factor K_I (MPa$\sqrt{m}$)	
(kN)	Finite Element	Equation 1
14.08	9.89	9.90
28.07	19.80	19.78
41.77	29.80	29.45
55.40	39.80	39.05
69.06	49.88	48.68

The stress intensity factor, K_I, was determined using the
stiffness derivative technique of Parks (8). The results of the
analysis are summarized in Table 2. For comparison, the stress
intensity factor calculated at the same loads using equation (1)
are also tabulated. The two results agree surprisingly well.

6. DISCUSSION

The good agreement between the finite element results and K_I values determined from equation (1) suggest that the grips constrain the curved centre-cracked specimen so that it behaves as a flat plate specimen. If this is correct it suggests that like the curved compact-tension specimen, the curved centre-cracked specimen can make use of analyses which already exist for the flat specimen in both the elastic and elastic-plastic regimes. However the finite element results must still be confirmed by an experimental compliance calibration.

There does not appear to be any noticeable effect on specimen behaviour due to friction between the grips and specimen. The stresses and strains produced in the specimen are primarily due to the increase in radius which results from the grip rotation. These stresses and strains occur regardless of any sliding between the grip surface and the specimen. In fact, circumferential sliding at a point will result in even greater radial deformation. Thus any variation in the stress distribution of the specimen which may be induced by frictional forces should be insignificant in comparison to the stresses induced by the radial (normal) forces. The ratio of frictional to normal forces is equal to the coefficient of friction, μ. With graphite as a lubricant, μ might equal 0.1, so the frictional forces will be only 10% of the normal forces and will not exert a noticeable effect.

7. CONCLUSIONS

A preliminary finite element analysis has indicated that, with a specially designed set of testing grips, fracture toughness tests can be performed on curved pressure tube specimens and the results analyzed using relations derived for flat plates. This technique is most promising for testing irradiated tube, but since no residual stresses are introduced from flattening and no stress relieving operations are required, this technique could potentially be extended to fracture toughness testing on tubular and piping materials in general. However, more development is required to confirm the analytical results and experimentally demonstrate the specimen's effectiveness.

REFERENCES

1. Langford, W.J. and L.E.J. Mooder. Fracture Behaviour of Zirconium Alloy Pressure Tubes. International Journal of Pressure Vessels and Piping, 6 (1979) pp. 275-310.
2. Simpson, L.A. and B.J.S. Wilkins. Prediction of Fast Fracture in Zr-2.5 wt% Nb Pressure Tubes Using Elastic-Plastic Fracture Mechanics, from Mechanical Behaviour of Materials, Proc. ICM3 (3), Cambridge, England, August 1979, Miller, K.J. and Smith, R.D. Eds., (Pergamon Press, 1979).
3. Pletka, B.J., E.R. Fuller and B.G. Loepke. An Evaluation of Double Torsion Testing - Experimental, from Fracture Mechanics Applied to Brittle Materials, ASTM STP 678, (American Society for Testing and Materials, 1979) pp. 19-57.
4. Davies, P.H. Curved Double Torsion Fracture Toughness Tests in Zr-2.5 wt% Nb Pressure Tube Material, Ontario Hydro Research Division Report No. M83-56-K, April 1983.
5. Simpson, L.A. The Application of Ductile-Fracture Analysis to Predictions of Pressure Tube Failure. Atomic Energy of Canada Ltd. Report No. AECL-6805, August 1981.
6. Simpson, L.A. Effects of Specimen Geometry on Elastic-Plastic R-Curves for Zr-2.5 wt% Nb, from Advances in Fracture Research, D. Francois et. al., Ed., pp. 833-841 (Pergamon Press, 1980).
7. Brown, W..F. and J.E. Srawley. Plane Strain Crack Toughness Testing of High Strength Metallic Materials, ASTM STP 410 (American Society for Testing and Materials, 1967), pp. 10-11.
8. Parks, D.M. A Stiffness Derivative Finite Element Technique for Determination of Elastic Crack Tip Stress Intensity Factors, International Journal of Fracture, 10 (1974), pp. 487-502.

ELEVENTH CANADIAN FRACTURE CONFERENCE ON TIME DEPENDENT FRACTURE

LIST OF PARTICIPANTS

Aifantis, E.C.
Dept. of Mech. Eng. & Engineering
 Mechanics
Michigan Technological University
Houghton, MI 49931

Argon, A.
Massachusetts Inst. of Technology
Cambridge, MA 02139

J.P. Bailon
Dept. de génie métallurgique
Ecole Polytechnique
Montréal, Québec H3C 3A7

Berkovits, A.
Dept. of Materials Eng.
Technion - I.I.T.
Haifa 32000   Israel

S.B. Biner
CANMET-PNRL
Energy, Mines & Resources Canada
Ottawa, Ontario K1A 0G1

Boutin, J.
CANMET
Energy, Mines & Resources Canada
Ottawa, Ont. K1A 0G1

Bratina, J.
Dept. of Metallurgy & Mat. Science
University of Toronto
184 College Street
Toronto, Ont. M5S 1A4

Burger, G.
McMaster University
Hamilton, Ont. L8S 4M1

Burns, D.J.
Dept. of Mechanical Engineering
University of Waterloo
Waterloo, Ont. N2L 3G1

Dickson, J.T.
Dept. de génie métallurgique
Ecole Polytechnique
Montréal, Québec.   H3C 3A7

Donovan, J.A.
Mechanical Engr.
University of Massachusetts
Amherst, MA 01003
USA

Ellyin, F.
Dept. of Mechanical Engineering
University of Alberta
Edmonton, Alta.   T6G 2G8

Epstein, J.S.
ESM - VPISU
Blacksburg, VA 24061
USA

Falkoff, M.Q.
Bell Laboratories
Whippany. Rd., 20-207
Whippany, NJ 07981

Faucher, B.
Energy, Mines & Resources
CANMET
Ottawa, Ont.   K1A 0G1

Febres, A.
Ecole Polytechnique
Montréal, Québec.   H3C 3A7

Fine, M.E.
Walter P. Murphy Prof. of Mat Science
Ethnological Institute
North Western University
Evanston, III 60201
USA

Gibbon, W.M.
Materials Div.
Ontario Research Foundation
Sheridan Park Research Community
Mississauga, Ont. L5K 1B3

Gooch, D.J.
Central Electricity Res. Lab.
Kelvin Avenue
Leatherhead, Surrey
United Kingdom

Hahn, G.T.
Vanderbilt University
Dept. of Mech & Mat. Science
Nashville, TN 37235

Hoeppner, D.W.
The Cockburn Center for Eng. Design
Faculty of Applied Sci. & Engineering
University of Toronto
Toronto, Ont. M5S 1A4

Holt, R.T.
Structures & Materials Lab
NAE/NRC
Ottawa, Ont. K1A OR6

Islam, M.
Mech Engr.
National Research Council
Ottawa, Ont. K1A OR6

Koul, A.K.
Structures & Materials Lab.
Bldg. M-13A Montreal Road,
NAE - NRC
Ottawa, Ontario. K1A OR6

Krausz, A.S.
Dept. of Mechanical Engineering
University of Ottawa
Ottawa, Ont. K1N 6N5

Krausz, K.
Dept. of Mechanical Engineering
University of Ottawa
Ottawa, Ont. K1N 6N5

Lalonde, S.
Ecole Polytechnique
Montréal, Qué. H3C 3A7

Lam, P.W.K.
Dept. of Chemical Engineering &
   Applied Chemistry
University of Toronto
Toronto, Ont. M5S 1A4

Marchand, N.
Research Assistant
60 Wadsworth #3G
Cambridge, MA 02142
USA

Mshana, J.
Dept. Mechanical Engineering
University of Ottawa
Ottawa, Ont. K1N 6N5

Nadiv, S.
Dept. of Materials Eng.
Technion - I.I.T.
Haifa 32000   Israel

Neimitz, A.
Dept. of Mech. Eng. & Engineerin
   Mechanics
Michigan Technological Universit
Houghton, MI 49931

Nguyen-Duy, P.
IREQ
1800 Montée Ste Julie
Varennes, Québec. JOL 2P0

Olaosebikan, O.
ESM - VPISU
Blacksburg, VA 24061
USA

Pelloux, R.M.
Research Assistant
60 Wadsworth #3G
Cambridge, MA 02142

Piggott, M.R.
Dept. of Chemical Engineering &
  Applied Chemistry
University of Toronto
Toronto, Ont. M5S 1A4

Provan, J.W.
Mechanical Engineering Department
McGill University
817 Sherbrooke St. W.
Montreal, P.Q. H3A 2K6

Radhakrishnan, V.M.
Metallurgy Dept.,
Indian Institute of Technology
Madras- 600036 India.

Reed-Hill, R.E.
Dept. Materials Science & Engineering
University of Florida
Gainesville, FL. 32611

Rubin, C.A.
Vanderbilt University
Dept. of Mech. & At. Science
Nashville, TN 37235

Sadananda, K.
Code 6393, Themo. Mat. Branch
U.S. Naval Research Lab
Dept. of the Navy
Washington, D.C. 20375
USA

Shimizu, H.
Materials Div.
Ontario Research Foundation
Sheridan Park Research Community
Mississauga, Ont. L5K 1B3

Smith, C.W.
ESM - VPISU
Blacksburg, VA 24061 USA

Sullivan, J.
Materials Div.
Ontario Research Foundation
Sheridan Park Research Community
Mississauga, Ont. L5K 1B3

Thamburaj, R.
M-13 NAE
National Research Council
Ottawa, Ont. K1A OR6

Vasatis, I.
Dept. of Mat. Sci. & Eng.
1039 Mass Avenue # 1
Cambridge, MA 02138 USA

Vosikovski, O.
Dept. of Mechanical Engineering
University of Waterloo
Waterloo, Ont. N2L 3G1

Wallace, A.C.
Metallurgical Eng. Branch
Chalk River Nuclear Lab.
Chalk River, Ont. KOJ 1J0

Wallace, W.
Structures & Materials Lab.
NAE - NRC Bldg M-13
OTTAWA, Ont. K1A OR6

Watson, P.G.
Dept. Materials Science & Engineering
University of Florida
Gainesville, FL 32611

Weertman, J.R.
Ethonological Insitute
Northwestern University
Evanston, IL 60201

Wilkinson, D.S.
Inst. for Materials Res.
McMaster University
1280 Main St. West
Hamilton, Ont. L8S 4M1

Yoshimura, H.
Vanderbilt University
Dept. of Mech. & Mat. Science
Nashville, TN 37235

Yue, S.
Dept. of Matallurgy & Mat. Science
University of Toronto
184 College Street
Toronto, Ont. M5S 1A4

Zhai, Z.H.
Mech. Engrg. Dept.
McGill University
817 Sherbrooke St. W.
Montreal, P.Q. H3A 2K6

SUBJECT INDEX